Computational Music Science

Series editors

Guerino Mazzola
Moreno Andreatta

More information about this series at http://www.springer.com/series/8349

Emmanuel Amiot

Music Through Fourier Space

Discrete Fourier Transform in Music Theory

 Springer

Emmanuel Amiot
Laboratoire de Mathématiques et Physique
Université de Perpignan Via Domitia
Perpignan
France

ISSN 1868-0305   ISSN 1868-0313   (electronic)
Computational Music Science
ISBN 978-3-319-83323-1   ISBN 978-3-319-45581-5   (eBook)
DOI 10.1007/978-3-319-45581-5

Printed on acid-free paper

This Springer imprint is published by Springer Nature
The registered company is Springer International Publishing AG
The registered company address is: Gewerbestrasse 11, 6330 Cham, Switzerland

Introduction

This book is *not* about harmonics, analysis or synthesis of sound. It deals with harmonic analysis but in the abstract realm of musical structures: scales, chords, rhythms, etc. It was but recently discovered that this kind of analysis can be performed on such abstract objects, and furthermore the results carry impressively meaningful significance in terms of already well-known musical concepts. Indeed in the last decade, the Discrete Fourier Transform (DFT for short) of musical structures has come to the fore in several domains and appears to be one of the most promising tools available to researchers in music theory. The DFT of a set (say a pitch-class set) is a list of complex numbers, called Fourier coefficients. They can be seen alternatively as pairs of real numbers, or vectors in a plane; each coefficient provides decisive information about some musical dimensions of the pitch-class set in question.

For instance, the DFT of C♯EGB♭ is

$$(4, 0, 0, 0, 4e^{4i\pi/3}, 0, 0, 0, 4e^{2i\pi/3}, 0, 0, 0)$$

where all the 0's show the periodic character of the chord, the sizes of the non-nil coefficients mean that the chord divides the octave equally in four parts, and the angles $(2i\pi/3, 4i\pi/3)$ specify which of the three diminished sevenths we are looking at.

From David Lewin's very first paper (1959) and its revival by Ian Quinn (2005), it came to be known that the *magnitude of Fourier coefficients*, i.e. the length of these vectors, tells us much about the *shape* of a musical structure, be it a scale, chord, or (periodic) rhythm. More precisely, two objects whose Fourier coefficients have equal magnitude are *homometric*, i.e. they share the same interval distribution; this generalization of isometry was initially studied in crystallography. *Saliency*, i.e. a large size of some Fourier coefficients, characterises very special scales, such as the diatonic, pentatonic, whole-tone scales. On the other hand, flat distributions of these magnitudes can be shown to correspond with uniform intervallic distributions, showing that these magnitudes yield a very concrete and perceptible musical meaning. Furthermore, nil Fourier coefficients are highly organised and play a vital role in the theory of tilings of the line, better known as "rhythmic canons."

Finally, the cutting-edge research is currently focused on the other component of Fourier coefficients, their directions (called phases). These phases appear to model some aspects of tonal music with unforeseen accuracy. Most of these aspects can be extended from the discrete to the continuous domain, allowing the consideration of microtonal music or arbitrary pitch, and interesting links with voice-leading theory. This type of analysis can also be defined for ordered collections of non-discrete pitch classes, enabling, for instance, comparisons of tunings.

Historical Survey and Contents

Historically, the Discrete Fourier Transform appeared in D. Lewin's very first paper in 1959 [62]. Its mention at the very end of the paper was as discreet as possible, anticipating an outraged reaction at the introduction of "high-level" mathematics in a music journal – a reaction which duly occurred. The paper was devoted to the interesting new notion of the Intervallic Relationship between two pc-sets[1], and its main result was that retrieval of A knowing a fixed set B and IFunc(A,B) was possible, provided B did not fall into a hodgepodge of so-called special cases – actually just those cases when at least one of the Fourier coefficients of B is 0. These were the times when Milton Babbitt proved his famous hexachordal theorem, probably with young Lewin's help. As we will see, its expression in terms of Fourier coefficients allows one to surmise that the perception of missing notes (or accents, in a rhythm) completely defines the motif's intervallic structure. These questions, together with any relevant definitions and properties (with some modern solutions to Lewin's and others' problems), are studied in Chapter 1.

Lewin himself returned to this notion in some of his last papers [63], which influenced the brilliant PhD research of I. Quinn, who encountered DFT and especially large Fourier coefficients as characteristic features of the prominent points of his "landscape of chords" [72], see Fig. 4.1. Since he had voluntarily left aside for readers of the *Journal of Music Theory* the 'stultifying' mathematical work involved in the proof of one of his nicer results, connecting Maximally Even Sets and large Fourier coefficients, I did it in [10], along with a complete discussion of all maxima of Fourier coefficients of all pc-sets, which is summarised and extended in Chapter 4. Lacklustre Fourier coefficients, with none showing particular saliency, are also studied in that chapter.

Meanwhile, two apparently extraneous topics involved a number of researchers in using the very same notion of DFT: homometry which is covered in Chapter 2 (see the state of the art in [2, 64] and Tom Johnson's recent compositions *Intervals* or *Trichords et tetrachords*); and rhythmic canons in Chapter 3 – which are really algebraic decompositions of cyclic groups as direct sums of subsets. The latter can be used either in the domain of periodic rhythms or pitches modulo some 'octave,' and were first extensively studied by Dan Tudor Vuza [94][2], then connected to the general

---

[1] I use the modern concept, though the term 'pitch-class set' had not yet been coined at the time. IFunc(A,B) is the histogram of the different possible intervals from A to B.

[2] At the time, probably the only theorist to mention Lewin's use of DFT.

theory of tiling by [19, 17] and developed in numerous publications [8, 18, 73] which managed to interest some leading pure mathematician theorists in the field (Matolcsi, Kolountzakis, Szabó) in musical notions such as Vuza canons.[3]

There were also improbable cross-overs, like looking for algebraic decompositions of pc-collections (is a minor scale a sum and difference of major scales?) [13], or an incursion into paleo-musicology, quantifying a quality of temperaments in the search for the tuning favoured by J.S. Bach [16], which unexpectedly warranted the use of DFT.

Aware of the intrinsic value of DFT, several researchers commented on it, trying to extend it to continuous pitch-classes [25] and/or to connect its values to voice-leadings [89, 88]. These and other generalisations to continuous spaces are studied in Chapter 5. Another very original development is the study of all Fourier coefficients with a given index of all pc-sets [50], also oriented towards questions of voice-leadings. On the other hand, consideration of the profile of the DFT enables characterisation of pc-sets in diverse voices or regions of tonal and atonal pieces [98, 99] as we will see in Chapter 6, which takes up the dimension that Quinn had left aside, the *phase* (or direction) of Fourier coefficients. The position of pairs of phases (angles) on a torus was only recently introduced in [15] but has known tremendously interesting developments since, for early romantic music analysis [96, 97] but also atonal compositions [98, 99]. Published analyses involve Debussy, Schubert, Beethoven, Bartok, Satie, Stravinsky, Webern, and many others. Other developments include, for instance, comparison of intervals inside chromatic clusters in Łutoslawski and Carter, using DFT of pitches (not pitch classes) by Cliff Callender [25].

A Couple of Examples

I must insist on the fact that DFT analysis is no longer some abstract consideration, but is done on actual music: consider for instance Chopin's *Etude op. 10, N°5*, wherein the pentatonic (black keys) played by the right hand is a subset of G♭ major played by the left hand; but so are many other subsets (or oversets). I previously pointed out in [10] that, because the pentatonic and diatonic scales are complementary Maximally Even Sets, one is included in the other up to transposition (warranting the name 'Chopin's Theorem' for this property of ME sets); however, it is much more significant to observe that these two scales have *identical Fourier coefficients with odd indexes*[4], which reflects spectacularly their kinship (see Chapter 6 and Fig. 4.7). I cannot wait to exhibit another spectacular example of the 'unreasonable efficiency' of DFT: Jason Yust's discovery [98] that in Bartok's *String Quartet 4* (iv), the accompaniment concentrates its energy in the second Fourier component while this component vanishes for the melody, and conversely for the sixth component (associated with the whole-tone character). This is again vastly superior to classic

---

[3] The musical aspect lies in the idea that a listener does not hear any repetition either in the motif nor in the pattern of entries of a Vuza canon.

[4] The other coefficients, with even indexes, have the same magnitude, but different directions.

'Set-Theory' subset-relationships (parts of this analysis and others are reproduced in this book), cf. Fig. 0.1 (further commented on in Chapters 4 and 6).

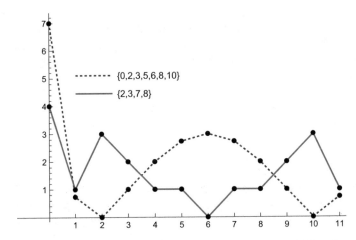

Fig. 0.1. DFT magnitudes of melody and accompaniment in Bartok

One explanation of the efficiency of DFT in music theory may well be Theorem 1.11. As we will see throughout this book, many music theory operations can be expressed in terms of *convolution products*. Not only is this product significantly simpler in Fourier space (i.e. after Fourier transform, cf. Theorem 1.10), but the aforementioned theorem proves that **Fourier space is the only one where such a simplification occurs**. This means that, for instance, interval functions or vectors, which are essential in the perception of the shape of musical objects, are more easily constructed and even perceived in Fourier space. *Idem* for the property of tiling – filling the space with one motif according to another – which is completely obvious when glancing at nil Fourier coefficients. Furthermore, we will see and understand how each and every polar coordinate in Fourier space carries rich musical meaning, not requiring any further computation.

Public

This book aims at being self-contained, providing coherent definitions and properties of DFT for the use of musicians (theorists and practitioners alike). A wealth of examples will also be given, and I have chosen the simplest ones since my purpose is clarity of exposition. More sophisticated examples can be found in the already abundant bibliography. I have also added a number of exercises, some with solutions, because the best way to make one's way through new notions is always with pen and pencil.

Professional musicians, researchers and teachers of music theory are of course the privileged public for this monograph. But I tried to make it accessible at pregraduate level, either in music or in mathematics. In the former case, besides introducing the notion of DFT itself for its intrinsic interest, it may help the student progress through useful mathematical concepts that crop up along the way. In the latter case, I hope that maths teachers may find interesting material for their classes, and that the musical angle can help enlighten those students who need a purpose before a concept. It is even hoped, and indeed expected, that hardened pure mathematicians will find in here a few original results worth their mettle.

Some general, elementary grounding in mathematics should be useful: knowledge of simple number sets (integers, rationals, real and complex numbers), basics of group theory (group structure, morphism, subgroups) which are mostly applied to the group $\mathbb{Z}_{12}$ of integers modulo 12; other simple quotient structures make furtive appearances in Chapters 1 and 3; vector spaces and diagonalization of matrixes are mentioned in Chapter 1 and used once in Chapter 2, providing sense to the otherwise mysterious 'rational spectral units'. The corresponding Theorem 2.10 is the only really difficult one in this book: many proofs are one-liners, most do not exceed paragraph length. All in all, I hope that any cultured reader with a smattering of scientific education will feel at ease with most of this book (and will be welcome to skip the remaining difficulties). On the other hand, mathematically minded but non-musician readers who cannot read musical scores or are unfamiliar with 'pc-sets' or 'scales' can rely on the omnipresent translations into mathematical terms.

Last but not least, some online content has been developed specifically for the readers of this book, who are strongly encouraged to use it: for instance all 'Fourier profiles' of all classes of pc-sets can be perused at

http://canonsrythmiques.free.fr/MaRecherche/photos-2/

while only a selection of the 210 cases is printed in Chapter 8, and software is available for the computation of the DFT of any pc-set in $\mathbb{Z}_{12}$.

Acknowledgements

First and foremost, my gratitude to Ian Quinn, who revived interest in DFT and invented the saliency quality. He is, even more than Lewin, simultaneously father and midwife of this new sub-discipline. May he be praised forever for this invaluable step forward.

I am very much indebted to Jason Yust, who made tremendous progress in the field in the last two years and generously gave me permission to cite all of his results and analyses, even those not yet published.

Jack Douthett is also father to a prolific notion, the Maximally Even Sets, which are a foundation to many further developments, including the present book. His support and encouragements were always a great help in my research.

I am grateful to Moreno Andreatta and Guerino Mazzola, who incited me to write this book and provided pointed and vital advice.

Among several memorable research collaborators or co-authors – Carlos Agon, Moreno Andreatta, Daniel Ghisi, John Mandereau, Thomas Noll – I would like to single out for the present opus William Sethares, since our joint work on matricial shortcuts through music-theoretical notions provided some major insights on the usefulness of DFT.

I have used many times a canon by composer Georges Bloch and I am grateful for his permission. Cliff Callender allowed me to borrow from the well-chosen examples of his paper on Fourier; his openness and helpfulness are as usual greatly appreciated.

Tom Johnson has been a constant source of stimulation in my research. He also provided many compositions of interest for this book, which would have been much more terse without him. He proves every day that these mathematical speculations of ours pave the way to making very real music.

I remember with pleasure the fruitful discussions on frequency of interval classes vs. Fourier saliency that were exchanged with Aline Honingh. It influenced the overall shaping of Chapter 4.

The music and maths community, organised around the SMCM conference and the *Journal of Mathematics and Music*, has been since its foundation a constant and stimulating source of inspiration. I would like to cite all its members.

IMSLP is an invaluable source for free-of-rights musical scores, from which I borrowed much more than appears in the printed book.

Many thanks to my proof-readers: my children Jeanne and Raphaël, Hélianthe Caure and especially lynx-eyed Serge Bastidas, who spotted obscure misprints even in the maths. Jeanne and my niece Cora were a great help in enhancing my home-made graphics. The professionalism of the Springer team and its readings is unrivaled.

Last but certainly not least, to my wife, Pascale, whose patience was several times severely tested during the elaboration of this book.

Saint Nazaire, P.O., France,   *Emmanuel Amiot*
October 2014 - May 2016.

Notations

- Sets are given between curly brackets: $\{0,4,7\}$. Sequences or n-uplets, taking into account the order of elements, use parentheses: $(0,7,4)$.
- $[a,b], [a,b[$ are respectively closed and semi-open intervals. For intervals of integers I use $[\![2,5]\!] = \{2,3,4,5\}$.
- $\mathbb{N}, \mathbb{Z}, \mathbb{Q}, \mathbb{R}, \mathbb{C}$ are respectively the sets of natural integers, integers, rationals, real and complex numbers.
- $a \mid b$ means that a is a divisor of b (most of the time a,b will be integers, in a few occasions I will use divisors of polynomials).
- $\mathrm{Div}(n)$ is the set of divisors of n: $\mathrm{Div}(12) = \{1,2,3,4,6,12\}$. The greatest common divisor is denoted by $\gcd(a,b)$.
- $\varphi(n)$ is Euler's totient function. Several definitions will be provided in this book.
- $|z|$ is the absolute value, or magnitude, of a real or complex number z.
- $\#A$ is the cardinality of the set A, i.e. its number of elements.
- $\mathbf{1}$ denotes the constant map with value 1. Any constant maps can thus be described as $c.\mathbf{1}$.
- $\mathbb{Z}_n$ is short for $\mathbb{Z}/n\mathbb{Z}$, the cyclic group (or ring) with n elements. 'Pitch-classes' (i.e. notes modulo octave equivalence) are modeled by the elements of $\mathbb{Z}_{12}$, or $\mathbb{Z}_n$ if the octave is divided into n parts. 'Pitch-class sets' or 'pc-sets' are subsets of $\mathbb{Z}_n$.
- More generally, the slash / denotes a quotient structure: $\mathbb{R}/(2\pi\mathbb{Z})$ means real numbers modulo any multiple of 2π, i.e. angles. In a few situations, more complex quotient structures are used (say $\mathbb{Z}[X]/(X^n - 1)$, i.e. a ring of polynomials modulo an ideal) and will be explained on the fly.
- T/I is the dihedral group (usually in $\mathbb{Z}_{12}$) whose elements are the transpositions (translations) $x \mapsto x + c$ and inversions (central symmetries) $x \mapsto -x + c$.
- Equality modulo some n is written $a \equiv b \pmod{n}$. In a few places, I will state polynomial equations modulo $X^n - 1$, meaning that all powers of X have their exponents reduced modulo n (e.g. $X^{3n+2} \equiv X^2$).
- Abbreviation 'iff' stands for 'if and only if', sometimes the symbol $\Longleftrightarrow$ will be used.
- The symbol $\approx$ is used for isomorphisms (ex: $\mathbb{Z}_3 \times \mathbb{Z}_4 \approx \mathbb{Z}_{12}$). It is also used for approximate values of numbers, without risk of confusion ($\pi \approx 3.14$).
- $A \setminus B$ is the set of elements of A which are not elements of B.
- $\mathbb{Z}_n^*$ is the multiplicative group of invertible elements in $\mathbb{Z}_n$, i.e. the classes of integers coprime with n. Unless n is prime, this is not to be confused with $\mathbb{Z}_n$ deprived of 0, i.e. $\mathbb{Z}_n \setminus \{0\}$: for instance, $\mathbb{Z}_{12}^* = \{1,5,7,11\}$. Similarly $K \setminus \{0\}$ means that 0 is omitted in set K.
- Direct products of structures ($\mathbb{Z}_3 \times \mathbb{Z}_4$) and direct sums ($\{0,3,6,9\} \oplus \{0,4,8\} = \mathbb{Z}_{12}$) will be used freely. There will be very few occurrences of *semi-direct products*, e.g. $\mathbb{Z}_{12} \rtimes \mathbb{Z}_2$, that the reader is welcome to skip if unfamiliar with this notion.
- In Chapter 6, I sometimes use the notation $t = 10, e = 11$ for readability.

Contents

1

Discrete Fourier Transform of Distributions

Summary. This chapter gives the basic definitions and tools for the DFT of subsets of a cyclic group, which can model for instance pitch-class sets or periodic rhythms. I introduce the ambient space of *distributions*, where pc-sets (or periodic rhythms) are the elements whose values are only 0's and 1's, and several important operations, most notably convolution which leads to 'multiplication d'accords' (transpositional combination), algebraic combinations of chords/scales, tiling, intervallic functions and many musical concepts.

Everything is defined and the chapter is hopefully self-contained, except perhaps Section 1.2.3 which uses some notions of linear algebra: eigenvalues of matrices and diagonalisation. Indeed it is hoped that the material in this chapter will be used for pedagogical purposes, as a motivation for studying complex numbers and exponentials, modular arithmetic, algebraic structures and so forth.

The important Theorem 1.11 proves that DFT is the only transform that simplifies the convolution product into the ordinary, termwise product.

1.1 Mathematical definitions and preliminary results

1.1.1 From pc-sets to an algebra of distributions

Many abstract musical objects can be seen as subsets of a cyclic group $\mathbb{Z}_n$. A chord, for instance, is a collection of elements of $\mathbb{Z}_{12}$, which models the 12 pitch-classes modulo octave. An equivalent definition is the map which associates to each pc its truth value: is it present or not? The same goes for a periodic rhythm measured against a set of (usually regular) elementary beats. This map is the characteristic map of the chord/rhythm, e.g. the C major triad $\{0, 4, 7\}$ can be seen as the map whose values on $\mathbb{Z}_{12} = \{0, 1, \ldots 11\}$ are $(1, 0, 0, 0, 1, 0, 0, 1, 0, 0, 0, 0)$ and a traditional tango or habanera rhythmic pattern is $(1, 0, 0, 1, 1, 0, 1, 0)$.

Some notions can be more finely defined as *distributions* on $\mathbb{Z}_n$, i.e. maps from $\mathbb{Z}_n \to \mathbb{R}$ or $\mathbb{C}$. To quote a single example, one can consider chords modulo octave and take into account the multiplicity of each pitch-class by associating to the chord the multiset of the number of occurrences of each pc. For instance, a full C major chord (C3 E3 G3 C4) yields after reduction modulo octave the multiset $(2, 0, 0, 0, 1, 0, 0, 1, 0, 0, 0, 0)$.

© Springer International Publishing Switzerland 2016
E. Amiot, *Music Through Fourier Space*, Computational Music Science,
DOI 10.1007/978-3-319-45581-5_1

Definition 1.1. *Let k be a field (usually $k = \mathbb{C}$). The set of distributions on $\mathbb{Z}_n$ is the k- vector set $k^{\mathbb{Z}_n}$ of maps from $\mathbb{Z}_n$ to k.*

The vector space structure is canonical: $k^{\mathbb{Z}_n}$ identifies with the space k^n of n-uplets by writing down the values taken by the distribution: $f = (f(0), f(1), \ldots f(n-1))$. Furthermore, this makes intuitive sense:

- Adding distributions generalizes enrichment of chords: as seen above, redoubling the root of a C major triad can be expressed as the addition

$$(1,0,0,0,1,0,0,1,0,0,0,0) + (1,0,0,0,0,0,0,0,0,0,0,0)$$
$$= (2,0,0,0,1,0,0,1,0,0,0,0)$$

- Conversely, negative values in a distribution are what is needed when removing an element:

$$(1,0,0,0,1,0,0,1,0,0,0,0) = (2,0,0,0,1,0,0,1,0,0,0,0)$$
$$+ (-1,0,0,0,0,0,0,0,0,0,0,0)$$

and they appear quite naturally in *operations* between distributions.
Here is a more spectacular example of an algebraic relationship between major and minor (harmonic) scales (cf. Fig. 1.9):
C minor = C major + E♭ major - F major.

$$(1,0,1,1,0,1,0,1,1,0,0,1) = (1,0,1,0,1,1,0,1,0,1,0,1)$$
$$+ (1,0,1,1,0,1,0,1,1,0,1,0)$$
$$- (1,0,1,0,1,1,0,1,0,1,1,0)$$

- Admittedly, though one can easily make a case for real values[1], *complex* values seem a bit extreme so far. As we will soon see, they are mandatory in Fourier spaces, and the two (real) dimensions of complex numbers both play essential musical roles.

The easiest way to enrich this vector space structure is *termwise multiplication*:

$$(f \times g)(t) = f(t) \times g(t) = (a_0, \ldots, a_{n-1}) \times (b_0, \ldots b_{n-1}) = (a_0 b_0, \ldots, a_{n-1} b_{n-1})$$

This is not devoid of musical interpretations (for characteristic functions, it is the $\cap$ operator which Yannis Xenakis heavily used in his sieving operations, for instance) but as we will see below, many musical situations can be modelised by a more complicated operation:

Definition 1.2. *The* convolution product *in $k^{\mathbb{Z}_n}$ is defined by*

$$\forall x \in \mathbb{Z}_n \quad f * g(x) = \sum_{t \in \mathbb{Z}_n} f(t) g(x - t)$$

---

[1] Say loudness of a pitch, or 'velocity' in MIDI format, or probabilities of occurrence in a score, to name a few possible meanings.

In practical terms it is computed between n-uplets, with indexes $0, 1 \ldots n - 1$ reduced modulo n if necessary.

Proposition 1.3. $*$ *is well-defined in* $k^{\mathbb{Z}_n}$, *associative and commutative. There is a neutral element (i.e. $\delta * f = f * \delta = f$ for all f), the Dirac distribution*

$$\delta : t \mapsto \begin{cases} 1 & \textit{if } t = 0 \\ 0 & \textit{else} \end{cases} \quad \textit{i.e. } \delta = (1, 0, 0, \ldots 0).$$

The vector space $k^{\mathbb{Z}_n}$ with operations $(+, ., *)$ is a k-algebra.

This last sentence means that all trivial but useful properties, such as distributivity of $*$ vs. $+$, etc., are satisfied.

Proof. Straightforward verification, left to the reader. Please notice that this operation works essentially because $\mathbb{Z}_n$ is an additive group.

This algebra is not a field, notably because it contains divisors of zero, which play vital roles in musical problems as we will see throughout the book.

To substantiate the above claim that convolution is an essential musical operation, we may mention straightaway Pierre Boulez's *multiplication d'accords*: 'multiplying', say, the G♯ minor triad by the minor second B-C is exactly the same as the convolution of their respective distributions as can be seen in Fig. 1.1[2]

$$(0,0,0,1,0,0,0,0,1,0,0,1)*(1,0,0,0,0,0,0,0,0,0,0,1)=(0,0,1,1,0,0,0,1,1,0,1,1)$$

$$\text{i.e. } \{3,8,11\} * \{-1,0\} = \{2,3,7,8,10,11\}$$

Fig. 1.1. down semitone $*$ G♯ minor triad

Systematic use of grace notes, such as happens in baroque or blues music for instance, brings in the same product with $\{11,0\}$. Other examples will be studied later in the book. A truly impressive one is [97] where Yust analyses the DFT of products of dyads in Webern's *Satz für Streichquartett* op. 5, n° 4.

---

[2] This is known in the U.S. as Transpositional Combination, see [34].

For the moment, let us alleviate notations with a customary simplification: a map is usually (though not always) better visualised as the sequence of its values on the ordered elements of the cyclic group $\mathbb{Z}_n$, i.e. we identify f with $(f(0), f(1), \ldots f(n-1))$ though of course one could use just any ordering of $\mathbb{Z}_n$, say $(f(1), f(2), \ldots f(n) = f(0))$. This slight abuse can be vindicated as an isomorphism

$$f \mapsto (f(0), f(1), \ldots f(n-1))$$

between maps (in $\mathbb{C}^{\mathbb{Z}_n}$) and n-vectors (in $\mathbb{C}^n$).

Furthermore, as the reader may have already noticed, we will denote identically integers and their classes modulo n since the context (hopefully) always makes clear whether the computation ought to be taken modulo n or not.

1.1.2 Introducing the Fourier transform

The most basic component of the Fourier transform is the exponential map $t \mapsto e^{it}$. Recall that (for real-valued argument t) the complex number e^{it} can be written with real and imaginary parts $\cos t + i \sin t$, and hence $|e^{it}| = 1$, i.e. e^{it} lies on the unit circle. Moreover, the map is 2π-periodic, $e^{i(t+2\pi)} = e^{it}$. This is the most obvious reason why the Discrete Fourier Transform will be so natural an instrument in modelising pitch-*classes*, meaning notes modulo octave: if for instance say the pitch-class (pc) for D is 2, this is identified with D an octave higher, i.e. $2 + 12 = 14$. Pcs are integers *modulo 12*, and hence the quantity $e^{2i\pi t/12}$ or, more generally, $e^{2i\pi kt/12}$ with k integer, is well-defined when t is a pitch-class.

Consider now a subset, say $S = \{0, 4, 8\} \subset \mathbb{Z}_{12}$, which can be seen as a representation of an augmented fifth. The associated distribution, its *characteristic function*, is the map

$$\mathbf{1}_S : \mathbb{Z}_{12} \to \mathbb{C}$$

which takes value 1 on the elements of S, and 0 elsewhere; the sequence of values of $\mathbf{1}_S$ on $\{0, 1, 2 \ldots 11\}$ is $(1, 0, 0, 0, 1, 0, 0, 0, 1, 0, 0, 0)$, and has period 4. Other maps share the same period, like $x \mapsto e^0 = 1, e^{i\pi x/2}, e^{i\pi x}, e^{3i\pi x/2}$ or $\cos(\pi x/2), \sin(3\pi x/2)$, and so on. Actually the mean-value of these four exponential maps is 0 for $x \in \mathbb{Z}_{12} \setminus S$ and 1 for $x \in S$. This can be seen in Fig. 1.2, where x takes real values between 0 and 12 and the real parts of the four exponentials and their mean is shown. The picture with the imaginary parts of these maps is similar.

The point of a Fourier transform is to express *any* map as a sum of complex exponentials. The relative importance of a given exponential component with period T tells how much the original map is T-periodic. Indeed the original aim of Fourier transform is the study of periodicities; or instance the characteristic map of an octatonic collection, with values $(1, 1, 0, 1, 1, 0, 1, 1, 0, 1, 1, 0)$, can be decomposed as

$$x \mapsto \frac{2}{3} + \frac{1}{6}\left((1 - i\sqrt{3})e^{2i\pi x/3} + (1 - i\sqrt{3})e^{4i\pi x/3} \right)$$

which is a combination of exponential maps with period 3.[3]

---

[3] For a very pedagogical explanation of the FT by and for music theorists, see [25]. This book aims at a higher mathematical level and is of necessity more terse in the basic definitions.

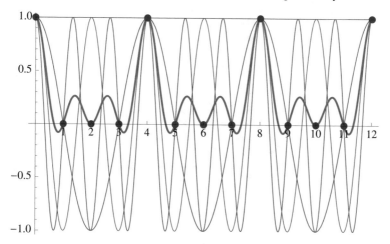

Fig. 1.2. The sum of four exponential functions

Fourier coefficients are *complex* numbers. As we will abundantly make clear below, the usual representation of complex numbers by real and imaginary part is inappropriate for musical applications and we will make use of the polar representation

$$z = x + iy = re^{i\varphi}$$

where $r = |z| = \sqrt{x^2 + y^2}$ is a positive number, the magnitude of z, and φ, the phase, is an angle (defined modulo 2π) characterised by the direction of z, i.e. $x = r\cos\varphi, y = r\sin\varphi$ (see Fig. 1.3). For instance the first coefficient for the diatonic collection C major, i.e. $(0, 2, 4, 5, 7, 9, 11)$, is

$$1 - \frac{\sqrt{3}}{2} + i\left(\frac{3}{2} - \sqrt{3}\right) = (2 - \sqrt{3})(\frac{1}{2} - i\frac{\sqrt{3}}{2}) = (2 - \sqrt{3})e^{-i\pi/3}.$$

Thousand of textbooks or webpages deal with various aspects and contexts of the Fourier transform and we refer the curious reader to this literature. The present text will be self-contained inasmuch as all the necessary notions are defined, alongside with some proofs.[4]

1.1.3 Basic notions

Definition 1.4. *The discrete Fourier transform (or DFT) of a distribution f on $\mathbb{Z}_n$, i.e. any map $f : \mathbb{Z}_n \to \mathbb{C}$, is another map from $\mathbb{Z}_n$ to $\mathbb{C}$ defined as follows:*

$$\widehat{f} : x \mapsto \sum_{k \in \mathbb{Z}_n} f(k)e^{-2i\pi kx/n}$$

---

[4] Though this book focuses on the *discrete* Fourier transform, we will devote some attention to continuous versions (see Chap. 5) which have appeared in the context of music theory. The continuous/integral Fourier transform (and Fourier series expansions) is of course well known in the theory of sound, but our topic is altogether different.

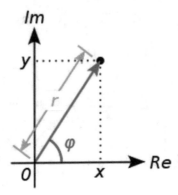

Fig. 1.3. The polar decomposition of a complex number

Alternatively, as we have seen above, f can be seen as the sequence of its values on the elements $(0, 1, 2 \ldots n-1)$ of $\mathbb{Z}_n$ and the Fourier transform acts on the vector set $\mathbb{C}^n$.

Please notice that, while in the formula k and x are integer **classes** mod. n and not genuine integers, the exponentials are nonetheless well defined (their values are the same whenever, say, x is replaced with $x \pm n$).

Theorem 1.5. *The DFT is a linear automorphism of $\mathbb{C}^{\mathbb{Z}_n} \approx \mathbb{C}^n$, the vector set of distributions on $\mathbb{Z}_n$. The reciprocal map $\widehat{f} \mapsto f$ can be explicitly written as the* **Fourier decomposition** *of f:*

$$f(k) = \frac{1}{n} \sum_{x \in \mathbb{Z}_n} \widehat{f}(x) e^{+2i\pi kx/n}$$

NB: There are alternative formulas and definitions of the Fourier and inverse Fourier transformation – for instance, a popular one for physicists uses the constant $1/\sqrt{n}$ in both cases, making the transform isometric for L^2 norm. This is of no consequence on what follows.

Proof. In this discrete case the proof of the formula is elementary:

$$\frac{1}{n} \sum_{x \in \mathbb{Z}_n} \widehat{f}(k) e^{+2i\pi kx/n} = \frac{1}{n} \sum_{x \in \mathbb{Z}_n} \sum_{\ell \in \mathbb{Z}_n} f(\ell) e^{-2i\pi \ell x/n} e^{+2i\pi kx/n}$$

$$= \frac{1}{n} \sum_{\ell \in \mathbb{Z}_n} \sum_{x \in \mathbb{Z}_n} f(\ell) e^{-2i\pi(\ell-k)x/n} = f(k)$$

since for $(\ell - k)x \neq 0$ (in $\mathbb{Z}_n$), $\sum_{x \in \mathbb{Z}_n} e^{-2i\pi(\ell-k)x/n}$ is nil as it is a sum of roots of unity according to Lemma 1.6:

Lemma 1.6. $\sum_{x \in \mathbb{Z}_n} e^{-2i\pi k/n} = 0$ *except when $k = 0$ modulo n.*

Proof.

$$(1 - e^{-2i\pi k/n}) \times \sum_{x=0}^{n-1} e^{-2i\pi kx/n}$$

$$= 1 - e^{-2i\pi k/n} + e^{-2i\pi k/n} - e^{-2i\pi 2k/n} + \cdots - e^{-2i\pi 2kn/n}$$

$$= 1 - 1 = 0.$$

Definition 1.7. *Values* $\widehat{f}(k)$ *for* $k = 0, 1, 2, \ldots n-1$ *are the* Fourier coefficients *of* f.

Theorem 1.8 (Parseval-Plancherel identity).

$$\sum_x |\widehat{f}(x)|^2 = n \sum_k |f(k)|^2$$

The proof is similar to the last one, reducing double sums by making good use of Lemma 1.6. This last theorem expresses a law of conservation of energy, or equivalently the preservation of the hilbertian norm. This is a very important (though often unnoticed) feature of the DFT: not only is it bijective (no loss of information), it is also isometric – energy-preserving. Some more trivial features of DFT are enumerated below without proof:

Proposition 1.9.

- *If* f *is real-valued then there is a skewed symmetry in the Fourier coefficients:* $\forall x \in \mathbb{Z}_n \ \widehat{f}(n-x) = \overline{\widehat{f}(x)}$ *(inverting* f *conjugates the DFT).*
- *If* f *has a period* p *(a divisor of* n*) then all Fourier coefficients are 0 except those whose index is a multiple of* n/p.[5]

As will appear in the musical applications throughout the book, the single most important operation on distributions is *convolution*, and this is the reason why DFT is so 'unreasonably efficient' in music theory.

Theorem 1.10 (convolution).
Recall the definition of the convolution product of two maps:

$$(f * g)(k) = \sum_{x \in \mathbb{Z}_n} f(k-x)g(x)$$

Then the DFT of a convolution product is the termwise *product of the DFTs of the maps:*

$$\forall x \in \mathbb{Z}_n \ \widehat{f * g}(x) = \widehat{f}(x) \times \widehat{g}(x)$$

Again this is easily proved, using only relabeling of sums on $\mathbb{Z}_n$. Theorem 1.10 goes a long way towards explaining the importance of DFT in music theory (and other fields): in effect, operating in Fourier space via termwise multiplication is analogous to filtering – when a coefficient of a distribution is nil, it will remain nil for any composition (by convolution) of this distribution with others. Guerino Mazzola

---

[5] This is easily understood as computing the DFT of the map induced by f by its restriction to $\mathbb{Z}/\frac{n}{p}\mathbb{Z}$. Again it follows from Lemma 1.6, see the exercises at the end of this chapter.

rightly questioned the existence of possible alternatives[6], but it so happens that DFT is unique in that respect:

Theorem 1.11. *The only linear automorphism of $\mathbb{C}^n$ that turns convolution $*$ to termwise product $\times$ is the DFT, up to permutation of the coefficients.*

Proof. Let G be such an automorphism, meaning that

$$G(f * g) = G(f) \times G(g)$$

for any distributions $f, g \in \mathbb{C}^n$. Let $f = \mathscr{F}^{-1}(\widehat{f}), g = \mathscr{F}^{-1}(\widehat{g})$ where F^{-1} is inverse DFT and $\widehat{f}, \widehat{g}$ are the DFTs of f, g. Then

$$\mathscr{F}(f * g) = \mathscr{F}(f) \times \mathscr{F}(g) = \widehat{f} \times \widehat{g} \iff f * g = \mathscr{F}^{-1}(\widehat{f} \times \widehat{g}),$$

meaning that

$$G(f * g) = G \circ \mathscr{F}^{-1}(\widehat{f} \times \widehat{g}) = G(f) \times G(g) = G \circ \mathscr{F}^{-1}(\widehat{f}) \times G \circ \mathscr{F}^{-1}(\widehat{g}).$$

Since $\widehat{f}, \widehat{g}$ are any distributions (because of the surjectivity of DFT) this means that $\Psi = G \circ \mathscr{F}^{-1}$ is not only a linear automorphism of $\mathbb{C}^n$, but also a multiplicative morphism.[7]

Consider the canonical vector basis $(e_1, \ldots e_n)$ where $e_1 = (1, 0, 0, \ldots), e_2 = (0, 1, 0, \ldots)$ and so on. Note that

$$e_i \times e_j = \begin{cases} (0, 0, \ldots 0) & \text{if } i \neq j \\ e_i & \text{if } i = j \end{cases}.$$

But if x satisfies $x^2 = x \times x = x$, i.e. $(x_1^2, x_2^2, \ldots x_n^2) = (x_1, x_2, \ldots x_n)$, then all the coordinates of x satisfy the same equation, hence for all $i, x_i = 0$ or 1. So all $\Psi(e_i)$ are 0-1 vectors, e.g. $(1, 0, 1, 1, 0, 0, 1 \ldots)$.

Furthermore, we have an orthogonality condition $\Psi(e_i) \times \Psi(e_j) = (0, 0, \ldots)$ for $i \neq j$ and this means that $\Psi(e_i)$ has no 1's in common with any other $\Psi(e_j), j \neq i$. Since the number of available places for 1's is n, equal to the number of vectors, it follows that each $\Psi(e_i)$ has one and exactly one 1. The positions of 1's being different, Ψ is a permutation of the e_i's.

Conversely, for any permutation $\sigma \in S_n$ of n indexes, we define a linear automorphism by: $\forall i, \Psi_\sigma(e_i) = e_{\sigma(i)}$, and Ψ_σ immediately commutes with $\times$ since it does so on a vector basis.

Finally we have $G \circ F^{-1} = \Psi_\sigma$ for some permutation σ, hence

$$G(f) = \Psi \circ F(f) = \Psi(\widehat{f}) = (\widehat{f}_{\sigma(0)}, \widehat{f}_{\sigma(1)}, \ldots).$$

---

[6] Modern harmonic analysis (in maths) uses many different orthogonal bases for decomposition of a signal, for instance wavelets; exponentials are only the seminal case. In calculus, the exponentials are privileged in being the eigenvectors of the differential operator, i.e. the simplest maps under differentiation.

[7] $(\mathbb{C}^n, \times)$ is not a group because many elements are not invertible. See Chapter 3.

An interesting cognitive consequence is that there is a significant reduction of complexity in thinking directly in Fourier space (inasmuch as we are doing something that would involve convolution in distribution/pc-sets space, anyway!), since termwise multiplication has complexity in n, whilst convolution is in n^2. For instance, checking that A tiles $\mathbb{Z}_n$ with B requires running through the whole of both A and B (quadratic time) in original space, because one has to check all pairs $(a, b) \in A \times B$ for their sum to run over the whole of $\mathbb{Z}_n$; however, this is done in linear time in Fourier space, checking once for each index whether the corresponding Fourier coefficient of A or B is nil. This appears at a glance on the Fourier graphs. The same goes for computing the interval function $\mathrm{IFunc}(A, B)$ of A, B.

1.2 DFT of subsets

1.2.1 What stems from the general definition

We define naturally:

Definition 1.12. *The DFT of a subset $A \subset \mathbb{Z}_n$ is the DFT of its characteristic function* $\mathbf{1}_A$:

$$\mathscr{F}_A = \widehat{\mathbf{1}_A} : x \mapsto \sum_{k \in A} e^{-2i\pi kx/n}$$

Though nowadays even pocket calculators allow computation in complex numbers, I provide the alternative cumbersome formulas using only real numbers (see also Section 3.3):

Proposition 1.13. *The DFT of a subset $A \subset \mathbb{Z}_n$ can alternatively be defined by its real and imaginary values:*

$$\mathfrak{R}(\mathscr{F}_A(x)) = \sum_{k \in A} \cos(2\pi kx/n) \qquad \mathfrak{I}(\mathscr{F}_A(x)) = -\sum_{k \in A} \sin(2\pi kx/n)$$

From there one can compute the all-important magnitude of $\mathscr{F}_A$ (see Chapters 2 to 5) as

$$|\mathscr{F}_A(x)| = \sqrt{\mathfrak{R}(\mathscr{F}_A(x))^2 + \mathfrak{I}(\mathscr{F}_A(x))^2}.$$

Notice that the DFT of any collection is a linear combination of DFTs of subsets or even single pcs, so that we are really studying a generating family in the vector space of distributions.

Example 1.14. The DFT of $\{0, 3, 6, 9\} \subset \mathbb{Z}_{12}$ is the map

$$x \mapsto \sum_{k=0}^{3} e^{-2i\pi 3kx/12} = 1 + (-i)^x + (-1)^x + i^x = \begin{cases} 4 & \text{if } x \in \{0, 4, 8\} \\ 0 & \text{else} \end{cases}.$$

Another way to put it is enumerating the Fourier coefficients: $(4,0,0,0,4,0,0,0,4,0,0,0)$, similar to those in Fig. 1.2. The real part of this map is a sum of four cosine functions, drawn for real values of x for dramatic effect, but the DFT exists only for $x \in \mathbb{Z}_{12}$, i.e. the blue dots (however see Chap. 5 for three meaningful extensions to continuous spaces).

For a more substantial example, recall Fig. 1.1 where Chopin uses the convolution product ('multiplication d'accords') of a dyad (descending semitone) and a triad. In Fig. 1.4 we can see three graphs representing the magnitudes of the DFT of the G♯ minor triad, the dyad, and their product the hexachord (F✗, G♯, A♯, B, C✗, D♯). We can observe that the third DFT (dashed line) is the product of the first two. Most notably, the value 0 for index 6 is inherited from the semitone, which hereby transmits its 'chromaticism' to the whole hexachord.

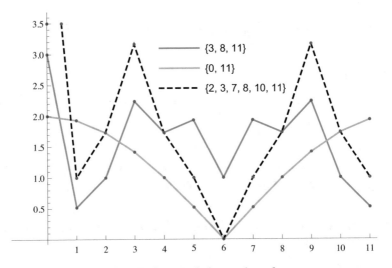

Fig. 1.4. DFT of a convolution product of pc-sets

The following results are elementary:

Proposition 1.15.

- *The DFT of a single note a is a single exponential function $t \mapsto e^{-2i\pi at/n}$, the DFT of the whole chromatic aggregate is*

$$\mathscr{F}_{\mathbb{Z}_n}(t) = \sum_{k=0}^{n-1} e^{-2i\pi kt/n} = 0 \qquad \text{for all } t \in \mathbb{Z}_n \setminus \{0\}, \mathscr{F}_{\mathbb{Z}_n}(0) = n.$$

- *Consequently, the Fourier transforms of a pc-set A and of its complement $\mathbb{Z}_n \setminus A$ have* opposite *values, except when $t = 0$:*

$$\forall t \in \mathbb{Z}_n \setminus \{0\}, \qquad \mathscr{F}_{\mathbb{Z}_n \setminus A}(t) = -\mathscr{F}_A(t).$$

- *The Fourier transform of A in 0 is equal to the cardinality of A: $\mathscr{F}_A(0) = \#A$.*
- *We have $A + p = A$ for some $0 < p < n$, meaning $A \subset \mathbb{Z}_n$ is periodic, if and only if $\mathscr{F}_A(t) = 0$ except when t belongs to the subgroup of $\mathbb{Z}_n$ with p elements.*[8]

This makes use of the linearity of the transform: the DFT of a pc-set is the sum of the DFTs of all pcs. This implies, by continuity, that small changes of the pc-set (say moving just one pc by one step) will change the DFT by a small amount (each coefficient moves by 1). More about relationships between voice-leading and DFT in Section 5.4.

Another easy computation yields the following key result:

Theorem 1.16. *The length, or magnitude, $|\mathscr{F}_A|$ of the Fourier transform*[9] *is invariant by (musical) transposition or inversion of the pc-set A. More precisely, for any $p,t \in \mathbb{Z}_n$*

- $\mathscr{F}_{A+p}(t) = \mathrm{e}^{-2\mathrm{i}p\pi t/n}\mathscr{F}_A(t)$ *(invariance under transposition)*
- $\mathscr{F}_{-A}(t) = \overline{\mathscr{F}_A(t)}$ *(invariance under inversion)*

Please do not confuse $-A$ (the inverse of A) and $-\mathbf{1}_A$ (negative pc-sets, equivalent to the complement of A under the equivalence in Proposition 1.18 below). The characteristic function of $-A$ is the *reverse* of $\mathbf{1}_A$, up to some circular permutation depending on the inversion chosen.

Example 1.17. Let $A = \{0,4,7\}$ in $\mathbb{Z}_{12}$ (C major triad). Then $11 - A = \{4,7,11\}$ or E minor triad. The respective characteristic functions are

$$\mathbf{1}_A = (\mathbf{1},0,0,0,\mathit{1},0,0,1,0,0,0,0) \qquad \mathbf{1}_{11-A} = (0,0,0,0,1,0,0,0,\mathit{1},0,0,0,\mathbf{1}).$$

Notice that some Fourier coefficients are invariant under some transpositions of the pc-set (whenever $p \times t$ is a multiple of n): for instance the fourth coefficient in equal 12-note temperament (henceforth 12 TeT) is invariant under minor third transposition ($t = 4, p = 3$).

It follows from the theorem that $|\mathscr{F}_A|$ is invariant under the T/I group of musical transformations[10], and even under complementation (except in 0 when $\#A \neq n/2$). As we will see and study in Chapter 2 about homometry, it is not a characteristic invariant (meaning $|\mathscr{F}_B|$ may be equal to $|\mathscr{F}_A|$ though A and B are not T/I related) because of the (in)famous Z-relation.

All the same, it appears to provide a very good snapshot of some relevant musical information of a given pc-set: by dropping the information of the *phase* of the Fourier coefficients and retaining only the *absolute value* or *magnitude*, we seem to focus on an essential part, the internal shape, in a way reminiscent of the Helmoltzian approach of sound perception, which showed that the phase of a sine wave can (in

---

[8] Generated in $\mathbb{Z}_n$ by n/p. Proof in exercises.

[9] This is a vector, listing the magnitudes of all coefficients.

[10] T/I is made of translations ('transpositions' for musicians) $x \mapsto x + \tau$ and central symmetries ('inversions') $x \mapsto c - x$.

many cases) be neglected, as the frequency is the part that enables recognition of pitch. Examples of Fourier magnitudes are given in Chapter 8, for instance major and minor triads are shown in Fig. 8.6. We will study the meaning of this magnitude in Chapter 4. The study of the meaning of phase is much more recent and it will be the topic of Chapter 6 of this book.

Another invariance occurs when one considers multisets instead of sets (this was first noticed and used by Yust in [98]):

Proposition 1.18. *The DFT does not change (except for its zeroth coefficient) when a constant is added to $\mathbf{1}_A$.*

Proof. Indeed $\mathscr{F}(\mathbf{1}_A + \lambda) = \mathscr{F}_A + \lambda \mathscr{F}(\mathbf{1}_{\mathbb{Z}_n})$ by linearity and $\mathscr{F}(\mathbf{1}_{\mathbb{Z}_n})$ is 0 except in 0, hence the result. $\square$

NB: $\mathscr{F}(\mathbf{1}_A + \lambda)(0) = \mathscr{F}_A(0) + n\lambda$, meaning that λ is added to the cardinality of each possible pc.

This allows us to make sense of non-positive values of a distribution: by adding a large enough constant to the distribution, one gets an equivalent distribution with positive quantities for each pc, changing only the cardinality of the multiset but no other Fourier coefficient.

There is a more complicated invariance result under affine maps. Recall that the affine transformations in $\mathbb{Z}_n$ are the maps $x \mapsto ax + b$ where $b \in \mathbb{Z}_n$ but $a \in \mathbb{Z}_n^*$, the group of invertibles in $\mathbb{Z}_n$ (which are the classes of the integers coprime with n). Multiplying a set by such an *invertible* element a is bijective, and *permutates* the Fourier coefficients:

Theorem 1.19.

$$\text{For all invertible } a \in \mathbb{Z}_n{}^*, \text{ for any } k \in \mathbb{Z}_n, \quad \mathscr{F}_{aA}(k) = \mathscr{F}_A(ak).$$

Hence affine maps preserve Fourier coefficients, but up to some permutation.

Proof.

$$\mathscr{F}_{aA}(x) = \sum_{k \in aA} e^{-2i\pi kx/n} = \sum_{k' \in A} e^{-2i\pi ak'x/n} = \mathscr{F}_A(ax).$$

Example 1.20. Here are the Fourier coefficients' magnitudes for pc-set $\{0, 1, 2\}$ and its multiple by 5, $\{0, 5, 10\}$:

$$(3, 1 + \sqrt{3}, 2, 1, 0, \sqrt{3} - 1, 1, \sqrt{3} - 1, 0, 1, 2, 1 + \sqrt{3}),$$
$$(3, \sqrt{3} - 1, 2, 1, 0, 1 + \sqrt{3}, 1, 1 + \sqrt{3}, 0, 1, 2, \sqrt{3} - 1).$$

Graphs for these lists – I call these *Fourier profiles* – appear in Fig. 8.9 and 8.7 in Section 8.3.

The most bizarre preservation is associated with oversampling. See Fig. 1.5 where we start with the simple motif (0,3,4,6) mod 8 (center): its repetition (left) is (0,3,4,6,8,11,12,14,16,19,20...) mod 32, while the other sampling changes the time unit and yields (right) (0, 12, 16, 24) mod 32.

Theorem 1.21. *Repeating a motif, i.e. turning $A = \{a_0, a_1, \dots\} \subset \mathbb{Z}_n$ into*

$$\overline{A_k} = \{a_0, a_1, \dots, a_0+n, a_1+n, \dots, a_0+2n, \dots a_0+(k-1)n, a_1+(k-1)n, \dots\}$$
$$= \{a_0, a_1, \dots\} \oplus \{0, n, 2n \dots (k-1)n\} \subset \mathbb{Z}_{kn},$$

is equivalent to oversampling *the DFT of A:*

$$\mathscr{F}_{\overline{A_k}} = k \times (\mathscr{F}_A(0), 0, 0, \dots, \mathscr{F}_A(1), 0, 0 \dots, \mathscr{F}_A(2), 0, 0 \dots) \in \mathbb{C}^{kn}$$

Conversely, oversampling a signal, i.e. turning $A = \{a_0, a_1, \dots\} \subset \mathbb{Z}_n$ into

$$B = \{a_0, 0, \dots, a_1, 0, \dots, a_2, 0, \dots\} \subset \mathbb{Z}_{kn}$$

is equivalent to repeating the Fourier transform k times:

$$\mathscr{F}_B = (\mathscr{F}_A(0), \mathscr{F}_A(1), \dots \mathscr{F}_A(n-1), \mathscr{F}_A(0), \mathscr{F}_A(1), \dots, \mathscr{F}_A(n-1),$$
$$\mathscr{F}_A(0), \mathscr{F}_A(1), \dots, \dots, \mathscr{F}_A(n-1))$$

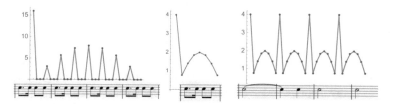

Fig. 1.5. Tango pattern in center, left repeated, right oversampled

Proof. One result implies the other by Inverse Fourier Transform. One or the other can be obtained by direct computation. Here we will more elegantly derive the first one from Theorem 1.10.

Let us denote $A = \{a_0, a_1, \dots, a_j \dots\} \subset \mathbb{Z}_n$ but A' for its copy $\{a_0, a_1, \dots\} \subset \mathbb{Z}_{kn}$. There are several different possible subsets A' actually, but this choice is irrelevant (this is discussed in more detail in Section 3.2.5). Now define

$$\overline{A_k} = \{a_0, a_1, \dots, a_0+n, a_1+n, \dots, a_0+2n, \dots a_0+(k-1)n, a_1+(k-1)n, \dots\} \text{ as}$$

$A' \oplus \{0, n, 2n \dots (k-1)n\}$ computed in $\mathbb{Z}_{kn}$. From the convolution formula, we know that

$$\mathscr{F}_{A' \oplus G} = \mathscr{F}_{A'} \times \mathscr{F}_G$$

The DFT of $G = \{0, n, 2n \dots (k-1)n\}$, in $\mathbb{Z}_{kn}$, is straightforward:

$$\mathscr{F}_G(t) = \sum_{j=0}^{k-1} e^{-2i\pi jnt/(kn)} = \sum_{j=0}^{k-1} e^{-2i\pi jt/k} = \begin{cases} \dfrac{1-e^{-2i\pi t}}{1-e^{-2i\pi t/k}} = 0 & \text{if } k \text{ does not divide } t \\ k & \text{when } k \mid t \end{cases}$$

Hence it only remains to check those coefficients of A' (not to confuse with A: we are computing the kn values of $\mathscr{F}_{A'}$ from the n values of $\mathscr{F}_A$) whose indexes are multiples of k, all others being 0:

$$\mathscr{F}_{A'}(kt) = \sum_j e^{-2i\pi a_j kt/(kn)} = \sum_j e^{-2i\pi a_j t/n} = \mathscr{F}_A(t), t \in \mathbb{Z}_n$$

and it follows that

$$\mathscr{F}_{\overline{A_k}}(t) = \begin{cases} 0 & \text{if } k \text{ does not divide } t \\ k\mathscr{F}_A(t') & \text{when } t = kt' \end{cases}$$

which is the expected oversampling of $\mathscr{F}_A$. $\square$

This means that, essentially, changing the frequency of sampling does not introduce new Fourier coefficients when the signal is already perfectly known. The point is that this enables one to compare pc-sets, or rhythms, originating in different universes (and with different numbers of beats/pitches). For instance, in Fig. 1.6 the clave rhythm $(0, 3, 6, 10, 12)$ has period 16, whereas the tango pattern $(0, 3, 4, 6)$ has period 8. Using some period multiple of both periods allows us to look at both DFTs simultaneously in a meaningful way. In Fig. 1.6, we compare rhythms with periods 4, 8 and 16 graphically[11] by the magnitudes of their DFTs in $\mathbb{Z}_{32}$, thus showing that the traditional tango pattern (upper right) is more similar to modern tango (tresilo, bottom left) or even traditional milonga (bottom right) than the clave (upper left). See also [13] on this topic. Similar analyses cab be conducted, for instance, on scales with different periodicities (say in different micro-tonal universes), avoiding the annoying *caveat* of many theories which are confined to same cardinalities.

In the domain of pitch-classes, the last theorem is of interest for the computation of DFT of Limited Transposition Modes, cf. the characteristic Figs. 8.31, 8.33, 8.30, 8.25.

1.2.2 Application to intervallic structure

We have seen that the magnitude of the DFT of a subset of $\mathbb{Z}_n$ (i.e. of the various Fourier coefficients) does not change when the subset is translated or inverted. This suggests a strong relationship between these magnitudes and the 'shape' of a subset. To be precise,

Definition 1.22. *The* interval function *between two subsets $A, B \subset \mathbb{Z}_n$ is the histogram of the intervals between A and B, i.e. $\mathrm{IFunc}(A,B)(k)$ is the number of pairs $(a,b) \in A \times B$ such that $b - a = k$.*

The interval content *of a subset $A \in \mathbb{Z}_n$ is*

$$\mathrm{IC}_A(k) = \mathrm{IFunc}(A,A)(k) = \#\{(i,j) \in A^2, \ i - j = k\}.$$

Example 1.23. In the first measures of Debussy's *Le vent dans la plaine* (Préludes, Livre I), the right hand plays a quick arpeggio on $B\flat C\flat$ while the left hand meanders around the three notes $D\flat, E\flat, G\flat$ (see Fig. 1.7).

---

[11] Quantitative correlations could be computed, of course, with any reasonable estimator.

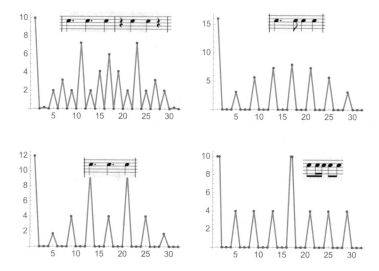

Fig. 1.6. Comparing four dance rhythms with different periods

Fig. 1.7. Two distinct pc-sets in *Le vent dans la plaine*

Let $A = \{10, 11\}$ and $B = \{1, 3, 6\}$, it is straightforward to compute $\mathrm{IFunc}(A, B) = (0, 0, 1, 1, 1, 1, 0, 1, 1, 0, 0, 0)$ – for instance, the initial 0 means that there are no common notes, and the fifth entry equal to $\mathrm{IFunc}(A, B)(4) = 1$ corresponds to $3 - 11 = 4$.

The only intervals inside B are primes, seconds, a minor third, a fourth and their reverses, hence $\mathrm{IC}(B) = (3, 0, 1, 1, 0, 1, 0, 1, 0, 1, 1, 0)$.

Lewin provided an appealing definition in his last papers on the subject: IC gives (up to a constant) a measure of the *probability of occurrence* of an interval between two notes in A (assuming all notes happen independently and with uniform probability). Notice that $\mathrm{IC}_A(0) = \#(A)$ and $\mathrm{IC}_A(n - k) = \mathrm{IC}_A(k)$, which makes at least half of the values redundant; these values are traditionally omitted in the interval vector **iv**. We need them all, though, in order to be able to compute DFT.

Theorem 1.24. *The knowledge of B and of* $\mathrm{IFunc}(A, B)$ *completely determines set A, except in 'Lewin's special cases' which can be summed up in the condition* $\prod_k \mathscr{F}_B(k) = 0$.

Theorem 1.25 (Lewin's Lemma).

The DFT of the intervallic content is equal to the square of the magnitude of the DFT of the set:

$$\forall k \in \mathbb{Z}_n \quad \widehat{IC_A}(k) = |\mathscr{F}_A(k)|^2.$$

Example: for $A = \{0,1,5\}$, the DFT is $(3, 1-i, 2, 1-2i, 0, 1-i, -1, 1+i, 0, 1+2i, 2, 1+i)$.

The IC is $(3,1,0,0,1,1,0,1,1,0,0,1)$ and its DFT is $(9,2,4,5,0,2,1,2,0,5,4,2)$, which is clearly equal to the square magnitudes of the complex DFT.

We will provide in Section 1.2.3 a more direct way to retrieve a pc-set A from B and IFunc(A,B).

Proof. Both theorems derive from the simple remark that IFunc is a convolution product:

$$\text{IFunc}(A,B)(k) = \#\{(a,b) \in A \times B, \ a-b = k\} = \sum_t \mathbf{1}_A(t)\mathbf{1}_B(t+k)$$

$$= \sum_t \mathbf{1}_{-A}(-t)\mathbf{1}_B(k-(-t))\sum_u \mathbf{1}_{-A}(u)\mathbf{1}_B(k-u) = \mathbf{1}_{-A} * \mathbf{1}_B(k).$$

But as we recalled earlier, the Fourier transform of a convolution product is the ordinary product of Fourier transforms, hence for two subsets:

$$\widehat{\text{IFunc}(A,B)} = \mathscr{F}_{-A} \times \mathscr{F}_B = \overline{\mathscr{F}_A} \times \mathscr{F}_B,$$

and $\mathscr{F}_A$ (which yields A by inverse Fourier transform) can be retrieved unless $\mathscr{F}_B$ vanishes; and when $B = A$,

$$\widehat{IC_A} = \widehat{\mathbf{1}_{-A}} \times \widehat{\mathbf{1}_A} = \overline{\mathscr{F}_A} \times \mathscr{F}_A = |\mathscr{F}_A|^2.$$

As a corollary, A is periodic iff IC_A is.

Note that the Fourier transform of any IC is a real positive-valued function, an uncommon occurrence among DFTs of integer-valued functions.[12]

This is a good moment for introducing the vexing question of the Z-relation, to be studied in depth in Chapter 2, and which can now be reformulated – or indeed defined – in DFT terms:

Definition 1.26. $A, B \subset \mathbb{Z}_n$ are homometric *if and only if they share the same interval content; or equivalently if the absolute values of their DFT are equal:*

$$AZB \iff IC_A = IC_B \iff |\mathscr{F}_A| = |\mathscr{F}_B|.$$

The equivalence stands because $|\mathscr{F}_A|$ holds all the information about IC_A by inverse Fourier transform. In Fortean tradition [42], A and B are Z-related when they are homometric but not isometric (i.e. not T/I related). Focusing on this more interesting

---

[12] The DFT of a real-valued function is non-real in general, it only satisfies $\hat{f}(-t) = \overline{\hat{f}(t)}$.

case is not suited to mathematical treatment, since this binary relation is not transitive, among other drawbacks.[13]

From there we also get a very short proof of the hexachord theorem, considered by some to be the first mathematically interesting result in music theory.

At the time he issued his first paper, David Lewin had come to work with Milton Babbitt, who was trying to prove the hexachord theorem:

Theorem 1.27. *If two hexachords (i.e. 6-note subsets of $\mathbb{Z}_{12}$) are complementary pc-sets in $\mathbb{Z}_{12}$, then they have the same intervallic content (same numbers of same intervals).*

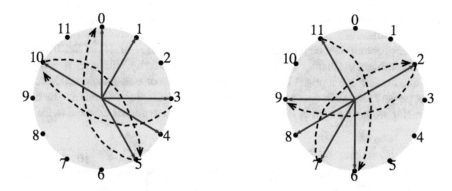

Fig. 1.8. These two hexachords have the same IC

In Fig. 1.8 with two complementary hexachords, the fifths have been signaled with curved arrows. Each hexachord has the same number of fifths, three in this example.

A simple derivation of this theorem in $\mathbb{Z}_n$ for any even n ensues from the trivial properties of DFT listed already:

Proof. As mentioned above, $\mathscr{F}_{\mathbb{Z}_n \setminus A} = -\mathscr{F}_A$, and this stands even for the 0th coefficient (the cardinality) since $A \in \mathbb{Z}_n$ has $n/2$ elements. So

$$\widehat{\mathrm{IC}_A} = |\mathscr{F}_A|^2 = |\mathscr{F}_{\mathbb{Z}_n \setminus A}|^2 = \widehat{\mathrm{IC}_{\mathbb{Z}_n \setminus A}}.$$

Hence (by inverse DFT) $\mathrm{IC}_A = \mathrm{IC}_{\mathbb{Z}_n \setminus A}$.

---

[13] This traditional position is not tenable; another argument against it is that some classes of 'Z-related' chords are indeed exchanged through action of a larger group than T/I, like the two famous all-intervals $\{0,1,4,6\}$ and $\{0,1,3,6\}$ in $\mathbb{Z}_{12}$ (Fig. 8.13), which are affine-related – and this can be generalised, since any affine transform of an all-interval set will be Z-related. Jon Wild pointed out to us that the reverse is false, and John Mandereau proved that the classes of homometric subsets are not orbits of any (point-wise) group action (Theorem 2.25), see [18] and Chapter 2 in this book.

More about generalisations of this theorem in Chapter 2 where we will also solve Lewin's 1959 original problem of pc-set retrieval, both via DFT and via a matrix formalism introduced in the next section. Notice the loss of information when we discard the direction of $\mathscr{F}_A$ and focus on its magnitude; it is the essence of the phase retrieval problem in Chapter 2.

1.2.3 Circulant matrixes

Several musical problems involve termwise-*division* of Fourier coefficients. Historically the first one is Lewin's problem [62]: retrieving a pc-set A from B and IFunc(A,B), say $B = \{1,3,6\}$ and IFunc$(A,B) = (0,0,1,1,1,1,0,1,1,0,0,0)$. As we have seen, this is essentially equivalent to the division of $\widehat{\text{IFunc}(A,B)}$ by $\mathscr{F}_B$. The reader is welcome to try it with pen and paper, without computing Fourier transforms of course.

Other problems can be reduced to the same mathematical solution. See the phase retrieval problem in Chapter 2. Another one was suggested by William Sethares: how does one decompose a pc-set into a linear combination of transpositions of another, given, pc-set? One such decomposition is C minor (harmonic) = C major + Eb major − F major, as can be seen in Fig. 1.9 (the red squares are pitches counted twice).

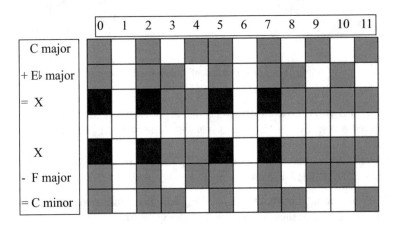

Fig. 1.9. Sum and difference of scales

This problem is very similar to the computation of IFunc, since it involves the convolution product of the characteristic function of a scale and the distribution of coefficients, respectively

$(1,0,1,0,1,1,0,1,0,1,0,1)$ and $(1,0,0,1,0,-1,0,0,0,0,0,0)$ in the above example. Remember our first example in Fig. 1.1, but here we have positive and negative coefficients (we will return to it in the study of direct sums in Section 3.2).

It was solved in [13] using the vector space of matrixes described below, which appears to be convenient for solving many similar problems.

Definition 1.28. *A circulating matrix is a square matrix whose coefficients move from one column to the next by circulating permutation, i.e. if the first column of S is* $(s_{0,0}, s_{1,0}, \ldots s_{n-1,0})^T$ *where* ^{T} *denotes (matricial!) transposition, then the next column is* $(s_{n-1,0}, s_{0,0}, s_{1,0}, \ldots s_{n-2,0})^T$ *and the* i^{th} *is* $(s_{n-i,0}, s_{n-i+1,0}, s_{1,0}, \ldots s_{n-i-1,0})^T$. *Equivalently the generic term is* $s_{i,j} = s_{i-j \pmod n},0 = s_{i-j \pmod n}$ *for short.*

Hence such a matrix is completely defined by its first column. We set $s_k = s_{k,0}$ for brevity, i.e. $S = \begin{pmatrix} s_0 & s_{n-1} & \cdots & s_1 \\ s_1 & s_0 & \cdots & s_2 \\ \vdots & & \ddots & \vdots \\ s_{n-1} & s_{n-2} & \cdots & s_0 \end{pmatrix}$.

The following results are standard linear algebra.

Theorem 1.29. *The set* $\mathscr{C}_n(k)$ *of all circulating matrixes with size n and coefficients in a field k is a sub-algebra[14] of* $\mathscr{M}_n(k)$. *It is commutative, its dimension is n.*

All matrixes in $\mathscr{C}_n(\mathbb{C})$ *can be simultaneously diagonalised, with eigenvectors*

$$y_m = \frac{1}{\sqrt{n}}\left(1, e^{-2i\pi m/n}, \ldots e^{-2i\pi m k/n}, \ldots, e^{-2i\pi m(n-1)/n}\right)^T \text{ for } m = 0, 1, \ldots n-1.$$

Set all these eigenvectors in matrix $\Omega = \frac{1}{\sqrt{n}}[e^{-2i\pi m k/n}]_{0 \leqslant k,m \leqslant n-1}$, *then* Ω *is a unitary matrix and for any* $S \in \mathscr{C}_n(\mathbb{C})$

$$\overline{\Omega}^T.S.\Omega = \Omega^{-1}.S.\Omega = \text{Diag}(\psi_0, \ldots \psi_{n-1})$$

where the eigenvalues of the diagonal matrix $\text{Diag}(\psi_0, \ldots \psi_{n-1})$ *are none other than the Fourier coefficients of the distribution* $(s_0, s_1, \ldots s_{n-1})^T$:

$$\psi_m = \sum_{k=0}^{n-1} s_k e^{-2i\pi k m/n}.$$

Proof. Any element of $\mathscr{C}_n(k)$ can be expressed as a polynomial in the matrix of the seminal circulating permutation:

$$S = s_0 I_n + s_1 J + s_2 J^2 + \ldots s_{n-1} J^{n-1} \qquad J = \begin{pmatrix} 0 & 0 & \cdots & 1 \\ 1 & 0 & \cdots & 0 \\ \vdots & \ddots & & \vdots \\ 0 & \cdots & 1 & 0 \end{pmatrix}.$$

It is straightforward to check that the columns of Ω are eigenvectors of the matrix J, with eigenvalue equal to the first element of the column. Hence

---

[14] An algebra is a vector space, with an internal multiplication. Common examples of algebras are square matrixes and polynomials. In this section we put forth several *algebra isomorphisms*, i.e. maps between algebras that preserve all three operations: addition, multiplication by numbers and internal multiplication.

$$\Omega^{-1}J\Omega = \begin{pmatrix} 1 & 0 & \cdots & 0 \\ 0 & e^{-2i\pi/n} & \cdots & 0 \\ \vdots & & \ddots & \vdots \\ 0 & \cdots & 0 & e^{-2i\pi(n-1)/n} \end{pmatrix}$$

and for $S = a_0I + a_1J + \cdots a_{n-1}J^{n-1}$,

$$\Omega^{-1}S\Omega = \begin{pmatrix} \psi_0 & 0 & \cdots & 0 \\ 0 & \psi_1 & \cdots & 0 \\ \vdots & & \ddots & \vdots \\ 0 & \cdots & 0 & \psi_{n-1} \end{pmatrix} \quad \text{where } \psi_k = \sum_{j=0}^{n} a_j e^{-2i\pi jk/n}.$$

This theorem proves that $\mathscr{C}_n(\mathbb{C})$ is isomorphic[15] with the sub-algebra $\mathscr{D}_n(\mathbb{C})$ of diagonal matrixes. Besides, it is trivially isomorphic with *the vector space* $\mathbb{C}^n$ (a circulating matrix is defined by its first column, which is an element of $\mathbb{C}^n$; and adding 2 such matrices is equivalent to adding the respective columns). So there must be an inner product in $\mathbb{C}^n$ which completes the isomorphism between algebras. This composition law is, of course, the convolution product of distributions

$$s*t = (\ldots, \sum_{k=0}^{n-1} s_{i-k}t_k, \ldots)^T.$$

We will study this law in yet another guise in Chapter 3, as the direct sum of subsets of $\mathbb{Z}_n$.

It is worthwhile to point out a direct isomorphism between the algebra of distributions $\mathbb{C}^n$ with the convolution product $*$ and the circulating matrixes algebra: any distribution f can be identified with the operator $\Phi_f = g \mapsto f*g$. This linear representation is bijective and the matrix of Φ_f in the canonical basis of $\mathbb{C}^n$ is the circulating matrix whose first column, the image of $e_0 = (1,0,0,0\ldots0)$, neutral for the convolution product, is the vector f itself.

From then we get an effective solution to Lewin's retrieval problem, which involves only the inversion of a matrix:

Proposition 1.30. *Let us define the matrix of a distribution s as the circulating matrix $\mathscr{S}$ whose first column is the set of values of s, and the* scale matrix *of a pc-set as the matrix $\mathscr{B}$ of the characteristic function b of B. Set $c =$ IFunc$(A,B) = \tilde{a}*b$ with associated matrix $\mathscr{C}$ where $\tilde{a}$ denotes the distribution of the* reverse *of a (i.e. the distribution of the opposite/inverse pc-set). Then*

$$\mathscr{C} = \mathscr{A}^T \times \mathscr{B}$$

and hence whenever $\mathscr{A}$ is invertible, $\mathscr{B} = \mathscr{C} \times \mathscr{A}^{T-1}$.

---

[15] Meaning that any operation in one structure is echoed by a similar operation in the other structure. The advantage is that matrix multiplication for diagonal matrixes is trivial, though it is not for ordinary matrixes, nor is convolution product of distributions. This is another expression of the simplification of convolution product to termwise multiplication, cf. Theorem 1.10.

Proof. The trick is that the matrix associated with the inverse subset $\tilde{a}$ is $\overline{\mathscr{A}^T} = \mathscr{A}^T$ which stands in fact for any real-valued a, which is the case studied in practice: this can be seen directly on the circulating matrix when reversing the first column, or by considering the eigenvalues with the symmetry $\psi_m = \overline{\psi}_{n-m}$. We stick to this case in the computations below.

One way to compute $\mathrm{IFunc}(A,B)$ is to calculate $\mathscr{A}^T \times \mathscr{B}$ and extract the first column. Indeed, if $\mathscr{A}$ and $\mathscr{B}$ are circulating matrixes for pc-sets a and b, then so is $\mathscr{C} = \mathscr{A}^T \mathscr{B}$,[16] and the kth element of the first (or rather, 0th) column of $\mathscr{C}$ is

$$\sum_{j=0}^{n-1} \mathscr{A}^T_{k,j} \mathscr{B}_{j,0} = \sum_{j=0}^{n-1} \mathscr{A}_{j,k} \mathscr{B}_{j,0} = \sum_{j=0}^{n-1} \begin{cases} 1 & \text{when } j-k \in A \text{ and } j-0 \in B \\ 0 & \text{else} \end{cases},$$

and the term in the sum is non zero only if $k = j - (j-k)$ is the distance between some element in a and another element in b: we recognise $\mathrm{IFunc}(A,B)$. Since $\mathscr{C}$ is a circulating matrix, the other columns are defined by this first column.

An alternative way makes use of special features of the matrixes, namely that $\overline{\mathscr{B}} = \mathscr{B}, \overline{\Omega}^{-1} = \Omega^T$ and for any circulating matrix S, the matrix with the Fourier coefficients $\Omega^{-1} S \Omega$ is diagonal. Hence, if we denote by $\mathscr{F}_A$ (resp. $\mathscr{F}_B$) the diagonal matrix of the Fourier coefficients of a (resp. b):

$$\mathscr{A}^T \mathscr{B} = \overline{\mathscr{A}}^T \mathscr{B} = \overline{\Omega \mathscr{F}_A \Omega^{-1}}^T \, \Omega \mathscr{F}_B \Omega^{-1} = \Omega \overline{\mathscr{F}_A}^T \Omega^{-1} \Omega \mathscr{F}_B \Omega^{-1} = \Omega \overline{\mathscr{F}_A} \mathscr{F}_B \Omega^{-1}$$

and we recognise the inverse Fourier transform of $\overline{\mathscr{F}_A} \mathscr{F}_B$, i.e. $\mathrm{IFunc}(A,B)$.

Example: Say that the matrix associated with $\mathrm{IFunc}(A,B)$ is

$$\mathscr{C} = \begin{pmatrix}
0 & 0 & 0 & 0 & 1 & 1 & 0 & 1 & 1 & 1 & 1 & 0 \\
0 & 0 & 0 & 0 & 0 & 1 & 1 & 0 & 1 & 1 & 1 & 1 \\
1 & 0 & 0 & 0 & 0 & 0 & 1 & 1 & 0 & 1 & 1 & 1 \\
1 & 1 & 0 & 0 & 0 & 0 & 0 & 1 & 1 & 0 & 1 & 1 \\
1 & 1 & 1 & 0 & 0 & 0 & 0 & 0 & 1 & 1 & 0 & 1 \\
1 & 1 & 1 & 1 & 0 & 0 & 0 & 0 & 0 & 1 & 1 & 0 \\
0 & 1 & 1 & 1 & 1 & 0 & 0 & 0 & 0 & 0 & 1 & 1 \\
1 & 0 & 1 & 1 & 1 & 1 & 0 & 0 & 0 & 0 & 0 & 1 \\
1 & 1 & 0 & 1 & 1 & 1 & 1 & 0 & 0 & 0 & 0 & 0 \\
0 & 1 & 1 & 0 & 1 & 1 & 1 & 1 & 0 & 0 & 0 & 0 \\
0 & 0 & 1 & 1 & 0 & 1 & 1 & 1 & 1 & 0 & 0 & 0 \\
0 & 0 & 0 & 1 & 1 & 0 & 1 & 1 & 1 & 1 & 0 & 0
\end{pmatrix},$$

---

[16] Because the algebra of circulating matrixes is stable under $\times$ and also under transposition of matrixes, since its generating element J satisfies $J^T = J^{-1} = J^{n-1}$.

$$\text{with } \mathscr{A} = \begin{pmatrix} 0\,1\,1\,0\,0\,0\,0\,0\,0\,0\,0\,0 \\ 0\,0\,1\,1\,0\,0\,0\,0\,0\,0\,0\,0 \\ 0\,0\,0\,1\,1\,0\,0\,0\,0\,0\,0\,0 \\ 0\,0\,0\,0\,1\,1\,0\,0\,0\,0\,0\,0 \\ 0\,0\,0\,0\,0\,1\,1\,0\,0\,0\,0\,0 \\ 0\,0\,0\,0\,0\,0\,1\,1\,0\,0\,0\,0 \\ 0\,0\,0\,0\,0\,0\,0\,1\,1\,0\,0\,0 \\ 0\,0\,0\,0\,0\,0\,0\,0\,1\,1\,0\,0 \\ 0\,0\,0\,0\,0\,0\,0\,0\,0\,1\,1\,0 \\ 0\,0\,0\,0\,0\,0\,0\,0\,0\,0\,1\,1 \\ 1\,0\,0\,0\,0\,0\,0\,0\,0\,0\,0\,1 \\ 1\,1\,0\,0\,0\,0\,0\,0\,0\,0\,0\,0 \end{pmatrix} ;$$

$$\text{Then } \mathscr{B} = \left(\mathscr{A}^{\mathscr{T}}\right)^{-1} \times \mathscr{C} = \begin{pmatrix} 0\,0\,0\,0\,0\,0\,1\,0\,0\,1\,0\,1 \\ 1\,0\,0\,0\,0\,0\,0\,1\,0\,0\,1\,0 \\ 0\,1\,0\,0\,0\,0\,0\,0\,1\,0\,0\,1 \\ 1\,0\,1\,0\,0\,0\,0\,0\,0\,1\,0\,0 \\ 0\,1\,0\,1\,0\,0\,0\,0\,0\,0\,1\,0 \\ 0\,0\,1\,0\,1\,0\,0\,0\,0\,0\,0\,1 \\ 1\,0\,0\,1\,0\,1\,0\,0\,0\,0\,0\,0 \\ 0\,1\,0\,0\,1\,0\,1\,0\,0\,0\,0\,0 \\ 0\,0\,1\,0\,0\,1\,0\,1\,0\,0\,0\,0 \\ 0\,0\,0\,1\,0\,0\,1\,0\,1\,0\,0\,0 \\ 0\,0\,0\,0\,1\,0\,0\,1\,0\,1\,0\,0 \\ 0\,0\,0\,0\,0\,1\,0\,0\,1\,0\,1\,0 \end{pmatrix}$$

i.e. $b = (0,1,0,1,0,0,1,0,0,0,0,0)$.[17]

For the time being, notice that this method works whenever $\mathscr{A}$ is invertible, which can be checked by computing its determinant instead of the Fourier coefficients. If $\mathscr{A}$ is singular, then other methods have been devised in [13] which are not relevant to this book.[18] Nil Fourier coefficients are actually a feature and not a hindrance in many situations, which will be considered in Section 3.1.

1.2.4 Polynomials

Partly for historical reasons, we mention yet another algebra isomorphic with the previous ones, $\mathbb{C}^n$ alias $\mathbb{C}^{\mathbb{Z}_n}$ and $\mathscr{C}_n(\mathbb{C})$ or $\mathscr{D}_n(\mathbb{C})$.

We have seen that any element of $\mathscr{C}_n(\mathbb{C})$ is a polynomial in the seminal circulating matrix J, $S = P(J)$ where $d^\circ P \leqslant n - 1$. On the other hand, for any polynomial

---

[17] We have thus retrieved the left hand (D♭E♭G♭) of *Le vent dans la plaine* from its interaction with the right hand. Notice that the computation of the first column of the matrix is enough. The bulk of the effort consists in inverting matrix $\mathscr{A}$, which can be done using techniques specific to $\mathscr{C}_n(k)$, though many pocket calculators have no trouble inverting 12×12 matrixes.

[18] See however Section 3.3.3 on algorithms, retrieving B even when one Fourier coefficient of A is nil.

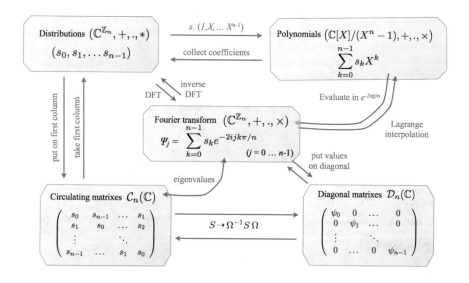

Fig. 1.10. Isomorphisms between algebras related to DFT

$P \in \mathbb{C}[X]$ the matrix $P(J)$ is in $\mathscr{C}_n(\mathbb{C})$, with eigenvalues $P(e^{-2mi\pi/n})$ associated with eigenvector y_m. Since $J^n = I_n$, P can be taken modulo $X^n - 1$ (i.e. $P(J)$ does not change when P is replaced with $P+$ some multiple of $X^n - 1$), and the induced map

$$P \in \mathbb{C}[X]/(X^n - 1) \mapsto P(J) \in \mathscr{C}_n(\mathbb{C})$$

whose domain is the quotient space $\mathbb{C}[X]$, modulo the multiples of $X^n - 1$, is an isomorphism[19](there is one and only one polynomial with degree $\leqslant n - 1$ taking the values ψ_m on the n points $e^{-2mi\pi/n}$, by polynomial interpolation). All the diverse isomorphisms are shown in Fig. 1.10. It is worthwhile to point out how a distribution – or its special case of a subset of $\mathbb{Z}_n$ – is associated[20] with a polynomial defined modulo $X^n - 1$. This transformation was apparently introduced by Redei in the study of tilings of $\mathbb{Z}_n$ in the 1950s.

Definition 1.31. *The* characteristic polynomial *of a set* $A \subset \mathbb{Z}_n$ *is*

$$\mathbf{A}(X) = \sum_{k \in A} X^k \in \mathbb{Z}[X]/(X^n - 1).$$

---

[19] I feel certain that some readers will prefer the shortcut of a short exact sequence

$$(X^n - 1) \hookrightarrow \mathbb{C}[X] \to \mathbb{C}[J].$$

[20] Meaning that X^n can be replaced by 1, and hence X^{n+1} is replaced by X, X^{5n+3} by X^3, etc.

More generally, for a distribution $s \in \mathbb{C}^n \approx \mathbb{C}^{\mathbb{Z}_n}$ one defines[21]

$$\mathbf{S}(X) = \sum_{k \in \mathbb{Z}_n} s(k) X^k \in \mathbb{Z}[X]/(X^n - 1).$$

The link with DFT is straightforward:

Proposition 1.32. *For any subset $A \subset \mathbb{Z}_n$ (or indeed for any distribution) we have*

$$\mathbf{A}(e^{-2ik\pi/n}) = \mathscr{F}_A(k).$$

This yields an isomorphism between Fourier space $\mathbb{C}^n$ and the vector space of polynomials with degree $< n$, or polynomials modulo $X^n - 1$, i.e. $\mathbb{C}[X]/(X^n - 1)$.

The map is obviously linear; it is bijective because a polynomial in $\mathbb{C}_{n-1}[X]$ is characterised by its values in n distinct points. Actually we have an algebra isomorphism:

Proposition 1.33.

- *convolution of distributions is expressed straightforwardly by the usual polynomial product, taken modulo $X^n - 1$; the characteristic polynomial of $c = a * b$ (a, b, c are distributions) is $\mathbf{C} = \mathbf{A} \times \mathbf{B} \mod (X^n - 1)$.*
- *Inversion of a pc-set A means taking the reciprocal polynomial*

$$\widetilde{\mathbf{A}}(X) = X^n \times \mathbf{A}(1/X).$$

Transposition by p is simply multiplication by X^p.

For instance the reciprocal polynomial of $1 + X^4 + X^7$ for $n = 12$ is $1 + X^8 + X^5$, which is a rather algebraic way of expressing that the C major triad reversed around C yields F minor.

Many results on tilings of the line (see Section 3.2) have been originally expressed or reached by way of polynomial computations, though the DFT definition developed below is more fashionable nowadays.

Example 1.34. Consider $S = (0, 7, 14, 21, 28, 35, 42)$, a diatonic scale generated by fifths in $\mathbb{Z}_{12}$; S can alternatively be written as $S = (0, 2, 4, 6, 7, 9, 11)$. The characteristic polynomial taken from this last expression is $\mathbf{S}(X) = 1 + X^2 + X^4 + X^6 + X^7 + X^9 + X^{11}$, but to appreciate its very special structure, one has to use the equivalent (modulo $X^{12} - 1$) formula

$$\mathbf{S}(X) = \sum_{k=0}^{6} X^{7k} = 1 + X^7 + \ldots X^{6 \times 7} = \frac{X^{49} - 1}{X^7 - 1}$$

which enables factoring $\mathbf{S}(X)$ in cyclotomic polynomial[22], more about which in Section 3.2.3.

---

[21] The exponents are defined modulo n since the polynomials are taken modulo $X^n - 1$.

[22] Here $\mathbf{S}(X)$ is actually equal to one lone cyclotomic polynomial, $\Phi_{49}(X)$.

NB: this polynomial should not be confused with the *characteristic polynomial of a matrix*, whose degree is always the size of the matrix. The characteristic polynomial of the circulating matrix of the diatonic scale above, for instance, is

$$\chi_D(X) = \det(XI_{12} - \mathscr{D}) = \prod_{k=0}^{12}(X - \mathscr{F}_D(k)) = \prod_{k=0}^{12}(X - S(e^{-2ik\pi/n}))$$
$$= X^{12} - 12X^{11} + 54X^{10} - 152X^9 + 138X^8 - 36X^7$$
$$+ 6X^6 + 12X^5 - 54X^4 + 152X^3 - 138X^2 + 36X - 7.$$

Exercises

Exercise 1.35. Check a few values for x in

$$x \mapsto \frac{2}{3} + \frac{1}{6}\left((1 - i\sqrt{3})e^{2i\pi x/3} + (1 - i\sqrt{3})e^{4i\pi x/3}\right)$$

(the result should be 1 unless $x - 2$ is a multiple of 3).

Exercise 1.36. Compute the DFT of $A = \{1, 4, 7, 10\}$. Using the fundamental $e^{ia+ib} = e^{ia} \times e^{ib}$, show that

$$\mathscr{F}_A(t) = e^{-it\pi/6}\mathscr{F}_B(t) \quad \text{where } B = \{0, 3, 6, 9\}.$$

Exercise 1.37. Prove Theorem 1.8.

Exercise 1.38. Check that $\forall x \in \mathbb{Z}_n \ \widehat{f}(n - x) = \overline{\widehat{f}(x)}$ (for a real-valued distribution f).

Exercise 1.39. Check the essential Theorem 1.10.

Exercise 1.40. Compute the convolution product of $\{1, 0, 0, 0, 0, 0, 0, 0, 0, 0, 0, 1\}$ and the minor triad $\{0, 0, 0, 1, 0, 0, 0, 0, 1, 0, 0, 1\}$ and multiply them. Compare with Fig. 1.1.

Exercise 1.41. Let p be a strict divisor of n. Prove that if $\mathscr{F}_A(x) = \mathscr{F}_{A+p}(x)$ and $\mathscr{F}_A(x) \neq 0$, then $e^{2i\pi px/n} = 1$, i.e. $\mathscr{F}_A(x) = 0$ unless x is a multiple of n/p.

Exercise 1.42. Compute by hand the third Fourier coefficients of $\{0, 4, 7\}$ and $\{3, 6, 10\}$ respectively and check that their magnitudes are equal.

Exercise 1.43. Compute the Fourier coefficients of pc-sets $\{0, 1, 2\}, \{0, 3, 6\}$ and $\{0, 5, 10\}$. Why are the middle ones so different from the two other cases?

Exercise 1.44. Compute the values of IFunc(A, B) when $A = \{0, 2, 7, 9\}$, $B = \{3, 5, 8, 10\}$, the interval contents of a minor triad and of Tristan's chord.

Exercise 1.45. Assuming IFunc$(A, B) = (0, 1, 0, 1, 0, 0, 1, 0, 1, 0, 0, 0)$ and $A = \{0, 2, 7, 9\}$, find B (hint: how many pcs in B ?).

Exercise 1.46. Pick up one hexachord, compute its interval content and do the same for its complement set.

Exercise 1.47. (Circulant matrixes) Write down the circulant matrixes for $\{0,11\}$ and the minor triad $\{3,8,11\}$ and multiply them. Compare with Fig. 1.1.

Exercise 1.48. (Polynomials) Write down the characteristic polynomials for $\{0,11\}$ and the minor triad $\{3,8,11\}$ and multiply them. Reduce the result modulo $X^{12} - 1$, i.e. transform any X^{k+12} into X^{k}. Compare with the previous exercise.

2

Homometry and the Phase Retrieval Problem

Summary. This chapter studies in depth the notion of homometry, i.e. having identical internal shape, as seen from Fourier space, where homometry can be seen at a glance by the size (or magnitude) of the Fourier coefficients. Finding homometric distributions is then a question of choosing the phases of these coefficients, hence this problem is often called *phase retrieval* in the literature. Such a choice of phases is summed up in the objects called *spectral units*, which connect homometric sets together. I included the original proof of the one difficult theorem of this book (Theorem 2.10), which non-mathematicians are quite welcome to skip. Some generalisations of the hexachord theorem are given, followed by the few easy results on higher-order homometry which deserve some room in this book because they rely heavily on DFT machinery. An original method for phase-retrieval with singular distributions (the difficult case) is also given. Some knowledge of basic linear algebra may help in this chapter.

We recall the definition of homometry and its characterisation given above: *Two subsets (or distributions) are homometric iff they share the same intervallic distribution, or equivalently iff they have the same magnitude for all their Fourier coefficients:*

$$\mathrm{IC}(A) = \mathrm{IC}(B) \iff |\mathscr{F}_A| = |\mathscr{F}_B|.$$

See the smallest example in Fig. 2.1 with $\{0,1,3,7\}, \{0,1,4,6\}$ and of course their retrogrades. Their intervallic function is $(4,1,1,1,1,1,2,1,1,1,1,1)$.

Though these pc-sets do appear in 20th century music (Elliot Carter's first quartet for instance), they had never been used as systematically as in Tom Johnson's *Intervals* (2013). The edges of the graph in Fig. 2.3 are the 48 homometric tetrachords, organised around common tritones for the eponymous piece. The composer navigates between adjacent tetrachords, each tritone being completed into the four distinct forms (up to transposition) of the tetrachords, as can be seen on the first line of the score in Fig. 2.2 (for instance 2,8 can be completed by 0,3 or 0,6 or 0,9 or 6,9). The other pieces in *Intervals*, seconds, thirds and so forth, similarly explore the same collection of 48 pc-sets with focus on seconds, minor thirds, etc –, since these tetrachords contain all possible intervals. The non-trivial homometry is clearly heard, the music spells the common tritones in the four different pc-sets classified in Fig. 2.1. Since the awakening of his interest in homometric sets, Johnson has worked on

© Springer International Publishing Switzerland 2016
E. Amiot, *Music Through Fourier Space*, Computational Music Science,
DOI 10.1007/978-3-319-45581-5_2

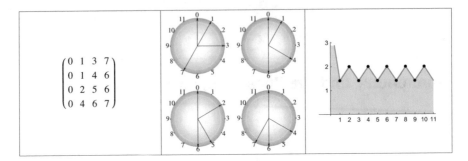

Fig. 2.1. Homometric quadruplets

other compositions using them, notably with homometric heptachords in the septet "Trichords et tetrachords" (2014).

Tritones

Fig. 2.2. Between common tritones

The Z-relation, as defined by Alan Forte [42] but also previously [49], is homometry in $\mathbb{Z}_{12}$, minus the trivial case of T/I related pc-sets. Non-trivially Z-related sets exist in $\mathbb{Z}_n$ for all $n \geq 12$, and also when $n = 8$ or 10.[1]

This notion originated in crystallography (see [75]) and addresses the question of whether an interference picture (say of a crystal under X-ray lighting) provides enough information to identify the geometric structure of the object.[2] The question of finding all (or at least some) homometric sets boils down to finding the *phase* of the Fourier coefficients, since their *magnitude* is common to all homometric distributions. Hence it is often called *the Phase Retrieval Problem* in the literature. Most of this chapter is adapted and simplified from [64, 2] which discuss the finer aspects of this problem and some possible generalisations, especially to non-discrete and/or non commutative groups. We will allude to some of these developments in Section 2.2.1.

---

[1] Examples in $\mathbb{Z}_n, n \geq 12$ are $A = \{0,1,2,6,8,11\}, B = \{0,1,6,7,9,11\}$.

[2] Interferences of lightwaves images are made by summing exponential waves with varying parameters, and the light intensity on the resulting picture varies proportionally to the magnitude of a Fourier transform.

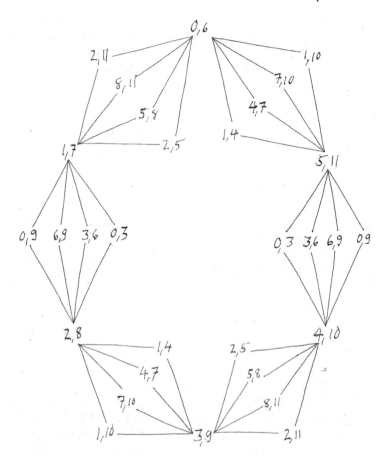

Fig. 2.3. T/I images of 0146 and 0137

2.1 Spectral units

In the most general setting, for a distribution f (recall that this generalizes the characteristic function of a pc-set, for instance) the intervallic function is the convolution product $d^2(f) = f * I(f)$, where $I(f)$ is the inversion of f, i.e. f read backwards, e.g. the traditional musical inversion in the case of a pc-set. This map is called the *Patterson function*; the notation d^2 probably means 'mutual *d*istances between points' for crystallographs. From Lewin's Lemma 1.25 we know that the Patterson function is completely determined by the Fourier transform, since

$$\widehat{d^2(f)} = |\widehat{f}|^2.$$

In this chapter, we will focus on the magnitude of the Fourier transform $|\widehat{f}|$ instead of $d^2(f)$. This is a simpler method, which would fail if $\mathbb{Z}_n$ were to be replaced by a non-commutative group since Fourier analysis is much more complicated in such contexts, but this still works in discrete and locally compact abelian groups as we will point out in Section 2.2.1. For the sake of simplicity, we stick to $\mathbb{Z}_n$ in the present section and refer the more curious readers to the bibliography.

2.1.1 Moving between two homometric distributions

Definition 2.1. *A distribution* $u \in \mathbb{C}^{\mathbb{Z}_n} \approx \mathbb{C}^n$ *is a* spectral unit *iff its Fourier transform is unimodular:*
$$\forall t \in \mathbb{Z}_n \ |\widehat{u}(t)| = 1.$$
We will denote the set of spectral units on $\mathbb{Z}_n$ *as* $\mathscr{U}_n$ *(or* $\mathscr{U}_n(K)$ *if we restrict ourselves to coefficients in a subfield* $K \subset \mathbb{C}$*).*

Theorem 2.2. *Two distributions* f, g *in* $\mathbb{C}^{\mathbb{Z}_n} \approx \mathbb{C}^n$ *are homometric iff there exists a spectral unit* u *such that* $f = u * g$.

Proof. This equation is equivalent to $\exists u \in \mathscr{U}_n, \ \widehat{f} = \widehat{u} \times \widehat{g}$ (termwise), which is itself equivalent to $|\widehat{f}| = 1 \times |\widehat{g}|$, i.e. to the homometry of f, g.

Proposition 2.3. $(\mathscr{U}_n, *)$ *is an abelian group.*

Proof. It is the image of the torus $((S^1)^n, \times)$ by inverse Fourier transform, which is a morphism as we have already established.

Since the Fourier transform is also an isometry (for a $\| \ \|_2$ norm, see Theorem 1.8), this means that the phase retrieval problem is solved and that the set of distributions homometric with f is simply $\mathscr{U}_n * f = \{u * f, u \in \mathscr{U}_n\}$, each orbit exhibiting the shape of a n-dimensional torus. This is true in a way, but deceptively, because we have chosen the smooth setting of the vector space $\mathbb{C}^n$ of *all* distributions. If one wishes to retrieve homometric **pc-sets**, then one must pick in the infinite orbit only those few distributions which are characteristic functions, i.e. with values in $\{0, 1\}$. Despite considerable efforts, there is still no known good general method for doing this. Computational methods are by and large the best practical tools but the complexity of their calculations is exponential.

Example 2.4. Consider again the most famous (non-trivial) homometric pc-sets in $\mathbb{Z}_{12}$, $A = \{0, 1, 4, 6\}$ and $B = \{0, 1, 3, 7\}$.

The spectral unit 'connecting' those two pc-sets, i.e. $1_B = 1_A * u$, is easily computed using techniques developed in the next . In this case, it is unique:
$$u = (\frac{1}{4}, \frac{1}{4}, 0, \frac{1}{4}, -\frac{1}{4}, -\frac{1}{2}, \frac{1}{4}, \frac{1}{4}, 0, \frac{1}{4}, -\frac{1}{4}, \frac{1}{2}).$$

Since one pc-set is the affine image of the other[3], the Fourier coefficients are actually permutated, following Theorem 1.19. The Fourier coefficients of u are unit-sized as expected:
$$\widehat{u} = \left(1, e^{\frac{i\pi}{6}}, e^{-\frac{i\pi}{3}}, i, 1, e^{\frac{5i\pi}{6}}, -1, e^{-\frac{5i\pi}{6}}, 1, -i, e^{\frac{i\pi}{3}}, e^{-\frac{i\pi}{6}}\right).$$

---

[3] In $\mathbb{Z}_{12}$, $\{0, 1, 3, 7\} = 5 \times \{0, 1, 4, 6\} - 5$.

2.1.2 Chosen spectral units

The fact that a distribution only takes values 0 or 1 yields *some* information about possible spectral units between this distribution and another homometric one; and we can refine Theorem 2.2:

Theorem 2.5. *If two pc-sets A, B are homometric then there exists a spectral unit u with rational coefficients such that $\mathbf{1}_A = u * \mathbf{1}_B$.*

This is an easy case of the more difficult following theorem ([76]):

Theorem 2.6 (Rosenblatt).
   Two distributions f, g in $\mathbb{Q}^{\mathbb{Z}_n} \approx \mathbb{Q}^n$ are homometric iff there exists a spectral unit u with values in $\mathbb{Q}$ such that $f = u * g$.

This statement is fairly obvious when one distribution is invertible for the convolution product $*$, since the coefficients of the inverse must stay in the same field (actually Rosenblatt's theorem is given for any subfield of $\mathbb{C}$). The difficulty is in the singular case (again Lewin's 'special cases'!) and we will see below that it remains so for higher-level homometry.

   Another simple case provides what is probably the most complicated proof of the hexachord theorem so far (no challenge intended):

Proposition 2.7. *Let $h = (\frac{2}{n} - 1, \frac{2}{n}, \frac{2}{n}, \dots)$. Then h is a spectral unit[4], and when n is even, for any set A with $n/2$ elements, $h * \mathbf{1}_A$ is equal to the characteristic function $\mathbf{1}_{\mathbb{Z}_n \setminus A}$ of the complement of A.*

Proof. Left as an exercise.

   The question of non-invertible, or singular, distributions, is studied in depth in Section 3.1. For the time being, it will suffice to make a simple observation: if f and g are homometric and $\widehat{f}(k) = 0$ (hence $\widehat{g}(k) = 0$ too) then the value of $\widehat{u}(k)$ can be chosen arbitrarily on the unit circle, with $f = u * g$ in k. In this case there may exist infinitely many different spectral units connecting f and g, even with restrictions on the coefficients. Some examples will be given below after we have developed the matricial technique for computation of spectral units.

   Remember the algebra of circulating matrixes $\mathscr{C}_n(\mathbb{C})$ in section 1.2.3. We have the following characterisation of spectral units in this setting (recall that the eigenvalues of the matrix are simply the Fourier coefficients of the distribution listed in its first column):

Proposition 2.8. *$u \in \mathbb{C}^n$ is a spectral unit $\iff$ its circulating matrix has all its eigenvalues on the unit circle. The group of such matrixes is the intersection of $\mathscr{C}_n(\mathbb{C})$ and the group of unitary matrixes (i.e. satisfying $U^{-1} = \overline{U}^T$).*

---

[4] Using the language of circulating matrixes which appear again *infra*, the matrix of h is
$$\mathscr{H} = \frac{2}{n} \begin{pmatrix} 1 & 1 & \dots & 1 \\ \vdots & \ddots & & \vdots \\ 1 & \dots & \dots & 1 \end{pmatrix} - I_n$$
and its eigenvalues are 1 and -1, this last repeated $n - 1$ times.

This makes even more obvious the isomorphism between $\mathcal{U}_n$ and the torus $\mathbb{T}_n$, which appears by diagonalisation. The whole group of rational (or real) spectral unit matrixes can be described implicitly by the equations

$$(E_k): \sum_{j=0}^{n-1} a_j a_{j+k} = 0, k = 1 \ldots \lfloor \frac{n-1}{2} \rfloor \text{ and } \sum_j a_j^2 = 1$$

where indices are taken modulo n. For instance, for $n = 3$ the group of *real* spectral units $\mathcal{U}_3(\mathbb{R})$ is the pair of parallel circles described by the matrixes $\begin{pmatrix} a & b & c \\ c & a & b \\ b & c & a \end{pmatrix}$ with $a^2 + b^2 + c^2 = 1$ and $a + b + c = \pm 1$.

Now the computation of a spectral unit between two given homometric distributions is straightforward:[5]

Proposition 2.9.

$$f = u * g \iff \mathscr{F} = \mathscr{U} \times \mathscr{G},$$

where $\mathscr{X}$ stands for the circulating matrix associated with distribution x.

In the example given above, we solved the equation in circulating matrixes (only the first column is provided)

$$\begin{pmatrix} 1 & \cdots \\ 1 & \cdots \\ 0 & \cdots \\ 1 & \cdots \\ 0 & \cdots \\ 0 & \cdots \\ 0 & \cdots \\ 1 & \cdots \\ 0 & \cdots \\ 0 & \cdots \\ 0 & \cdots \\ 0 & \cdots \end{pmatrix} = \mathscr{U} \times \begin{pmatrix} 1 & \cdots \\ 1 & \cdots \\ 0 & \cdots \\ 0 & \cdots \\ 1 & \cdots \\ 0 & \cdots \\ 1 & \cdots \\ 0 & \cdots \\ 0 & \cdots \\ 0 & \cdots \\ 0 & \cdots \\ 0 & \cdots \end{pmatrix}$$

which is done by inverting the right-hand matrix given.

2.1.3 Rational spectral units with finite order

Musical transposition is very simply and universally achieved by convolution with the spectral unit $j = (0, 1, 0, 0, 0, 0, 0, 0, 0, 0, 0, 0)$ and its powers, e.g. Eb minor triad is obtained from $\mathscr{S}$ by the matrix product $\mathscr{J}^3 \mathscr{S}$, or equivalently $j^3 * s = (0, 0, 0, 1, 0, 0, 1, 0, 0, 0, 1, 0)$. It is, however, much less straightforward to achieve *inversion* by way of a spectral unit.

Let $\mathscr{S}$ be the matrix of distribution $s = (1, 0, 0, 1, 0, 0, 0, 1, 0, 0, 0, 0)$ (the C minor triad) and $\mathscr{T}$ defined by $t = (1, 0, 0, 0, 1, 0, 0, 1, 0, 0, 0, 0)$ (the C major triad). From C major to C minor we must have $\mathscr{U} = \mathscr{S}^{-1} \mathscr{T}$, which yields

$$u = \frac{1}{15}(7, 4, -2, 1, 7, 4, -2, 1, -8, 4, -2, 1).$$

---

[5] With the proviso made above in the case of distributions with some nil Fourier coefficients, which can still be settled but via the arbitrary choice of the corresponding Fourier coefficients in $\hat{u}$. See Example 2.23.

Contrarily to transposition, the spectral unit achieving inversion *depends on the inverted subset*[6] (or distribution), and even more strangely, in general, such units are of infinite order in the group of units, as in the example above.

On the other hand, iterating convolution by the spectral unit connecting $\{0,1,3,7\}$ and $\{0,1,4,6\}$, which has finite order (all its Fourier coefficient are 12th roots of unity), yields twelve different distributions, eight of which are genuine pc-sets:

$$
\begin{bmatrix}
1 & 1 & 0 & 0 & 1 & 0 & 1 & 0 & 0 & 0 & 0 & 0 \\
1 & 1 & 0 & 1 & 0 & 0 & 0 & 1 & 0 & 0 & 0 & 0 \\
\frac{1}{2} & \frac{1}{2} & 1 & \frac{1}{2} & \frac{1}{2} & -1 & \frac{1}{2} & \frac{1}{2} & 0 & \frac{1}{2} & \frac{1}{2} & 0 \\
0 & 1 & 0 & 1 & 0 & 0 & 0 & 0 & 0 & 1 & 1 & 0 \\
1 & 0 & 0 & 0 & 1 & 0 & 0 & 0 & 0 & 1 & 1 & 0 \\
\frac{1}{2} & \frac{1}{2} & -1 & \frac{1}{2} & \frac{1}{2} & 0 & \frac{1}{2} & \frac{1}{2} & 0 & \frac{1}{2} & \frac{1}{2} & 1 \\
1 & 0 & 0 & 0 & 0 & 0 & 1 & 1 & 0 & 0 & 1 & 0 \\
0 & 1 & 0 & 0 & 0 & 0 & 1 & 1 & 0 & 1 & 0 & 0 \\
\frac{1}{2} & \frac{1}{2} & 0 & \frac{1}{2} & \frac{1}{2} & 0 & \frac{1}{2} & \frac{1}{2} & 1 & \frac{1}{2} & \frac{1}{2} & -1 \\
0 & 0 & 0 & 1 & 1 & 0 & 0 & 1 & 0 & 1 & 0 & 0 \\
0 & 0 & 0 & 1 & 1 & 0 & 1 & 0 & 0 & 0 & 1 & 0 \\
\frac{1}{2} & \frac{1}{2} & 0 & \frac{1}{2} & \frac{1}{2} & 1 & \frac{1}{2} & \frac{1}{2} & -1 & \frac{1}{2} & \frac{1}{2} & 0
\end{bmatrix}
$$

One can interpret the first four distributions in this table as splitting the minor third-down transposition into three identical moves, which are *not* transpositions (the first turns 0137 into 0146, and the next distribution is not a genuine pc-set), e.g. we have defined a non-trivial cubic root of the minor third transposition.

Since the study of rational spectral units with infinite order does not look too promising, it is natural to wonder about rational spectral units with **finite** order. Their set is a subgroup of $\mathscr{U}_n(\mathbb{Q})$. Since there are already, for instance, infinitely many matrixes 2×2 with rational coefficients and finite order, the following result is noteworthy. Moreover it is a practical way for exploring homometric classes in $\mathbb{Z}_n$, when n is not too large (though brute force search may seem more efficient, until more refined applications of this theorem are implemented).

Theorem 2.10. *Any spectral unit (represented as a rational circulating matrix) with finite order is completely determined by the values of the subset $\{\xi_j, j \mid n\}$ of its eigenvalues, the possibilities being listed* infra*:*

- $\xi_0 = \pm 1$;
- *When n is odd, for all $j \mid n$, ξ_j OR $-\xi_j$ is any power of $e^{2ij\pi/n}$.*
- *When n is even, ξ_j is any power of $e^{2ij\pi/n}$ if n/j is even, or any power of $e^{ij\pi/n}$ if n/j is odd.*

Then for any k coprime with n, $\xi_{kj} = \xi_j^k$ (or $-\xi_j^k$ in the specific case when ξ_j is a $e^{(2p_j+1)i\pi/n}$ and k is even).

As a corollary, we have the structure of the whole group:

---

[6] Matricially, one can write $\mathscr{U} = \mathscr{S}(\mathscr{S}^T)^{-1}$ if $\mathscr{S}$ is not singular.

Theorem 2.11. *The group of all rational spectral units with finite order in dimension n is isomorphic to the product of cyclic groups $\prod_{d|n} \mathbb{Z}_{\operatorname{lcm}(2,d)}$.*

These theorems may perhaps enable computation of all spectral units with, say, small denominators, which occur in practice for homometric subsets of $\mathbb{Z}_n$ and may be a provable condition in general cases.

For instance, for $n = 12$ the structure of this group is $\mathbb{Z}_{12} \times (\mathbb{Z}_6)^2 \times \mathbb{Z}_4 \times (\mathbb{Z}_2)^2$, with 6,912 elements. The denominators of the values of these spectral units are all divisors of 12, a typical one being

$$u = \left(-\frac{1}{12}, -\frac{1}{12}, 0, \frac{1}{4}, -\frac{1}{12}, -\frac{1}{3}, -\frac{3}{4}, \frac{1}{4}, -\frac{1}{3}, -\frac{1}{12}, \frac{1}{4}, 0\right).$$

The proof is quite involved, and non-mathematically inclined readers are invited to skip it.

Proof. We begin by proving two intermediary results, which are contained in the main theorem:

Lemma 2.12. *If $\mathscr{U} \in \mathscr{C}_n$ is a spectral unit (matrix) with finite order and n is even, then all its eigenvalues are n^{th} roots of unity. If n is odd, then the eigenvalues are either n^{th} roots of unity or their opposites (i.e. they are $2n^{th}$ roots of unity).*

This stems from a more precise condition:

Lemma 2.13. *If $\mathscr{U} \in \mathscr{C}_n$ is a rational spectral unit with finite order and n is even, then for all k coprime with n and any Fourier coefficient (= eigenvalue of $\mathscr{U}$) ξ_j, $j \neq 0$, one has $\xi_{kj} = \xi_j{}^k$. For $j = 0$ we have $\xi_0 = \pm 1$.*
The same condition stands when n is odd, with the exception of the case when ξ_j is a $e^{(2p_j+1)i\pi/n}$ and k is even: then $\xi_{kj} = -\xi_j{}^k$.

For instance for $k = -1$, this gives the condition that the last Fourier coefficients must be the conjugates of the first ones (thus ensuring that $\mathscr{U}$ is real valued). More generally, given one coefficient ξ_j we know all coefficients with indexes *associated* with j.

Throughout, $\mathscr{U}$ is a circulating matrix which is unitary ($\mathscr{U}^{-1} = {}^t\overline{\mathscr{U}}$), has finite order ($\mathscr{U}^m = I_n$ for some m), and has rational elements. Hence its eigenvalues have magnitude 1 (they are m^{th} roots of unity), and, as discussed above,xx $\mathscr{U}$ diagonalises into $\operatorname{Diag}(\xi_0, \xi_1, \ldots \xi_{n-1})$ where the eigenvalues ξ_j are also the Fourier coefficients of the first column of $\mathscr{U}$, seen as a map from $\mathbb{Z}_n$ to $\mathbb{C}$.

We begin by proving an alternative, simpler form of Lemma 2.12, stating that $m = n$ or $m = 2n$:

Lemma 2.14. *All eigenvalues of $\mathscr{U}$ are n^{th} roots of unity for even n, and $2n^{th}$ roots of unity for odd n.*

Already this establishes that the group we are looking for is finite, a non-trivial fact.

Proof. As we assumed that $\mathscr{U}$ has finite order, all its eigenvalues are roots of unity. Moreover, as $\mathscr{U} = P(\mathscr{J}), P \in \mathbb{Q}[X]$ is a polynomial in the matrix $\mathscr{J}$, whose eigenvalues are the nth roots of unity, the eigenvalues of $\mathscr{U}$ are polynomials in these roots, i.e. $\xi_k = P(e^{2ik\pi/n})$, and hence lie in the cyclotomic field $\mathbb{Q}_n = \mathbb{Q}[e^{2i\pi/n}]$. We need the following:

Lemma 2.15. *Let ξ be a mth root of unity belonging to the cyclotomic field $\mathbb{Q}_n$.*

$$\text{Then} \begin{cases} \xi^n = 1 & \text{when } n \text{ is even,} \\ \xi^{2n} = 1 & \text{when } n \text{ is odd.} \end{cases}$$

In other words, if $\mathbb{Q}_m \subset \mathbb{Q}_n$ then m is a divisor of n or $2n$, according to whether n is even or odd.[7]

Let ξ be such a unit root (say, any eigenvalue of $\mathscr{U}$). Let m be the order of ξ, i.e. the smallest integer satisfying $\xi^m = 1$; we know that ξ, primitive root of order m, generates $\mathbb{Q}_m$. As $\xi \in \mathbb{Q}[e^{2i\pi/n}]$ too, $\mathbb{Q}_m \subset \mathbb{Q}_n$. This does not obviously preclude $m > n$. We need still another:

Lemma 2.16. *The multiplicative group of elements of finite order in $\mathbb{Q}_n$ is cyclic.*[8]

This is because given two elements ξ, ξ' with orders m, m' it is possible to construct an element of order $\text{lcm}(m, m')$ (for instance, their product). In other words, the roots of unity in $\mathbb{Q}_n$ have a maximum order, which is the lcm of all possible orders.

Let us call again m this maximal value; to prove Lemma 2.12 we need to prove that $m = n$ or $2n$. Now, any element ξ of $\mathbb{Q}_n$ which is a root of unity must satisfy $\xi^m = 1$.

This is true in particular when ξ is the primitive n^{th} root $e^{2i\pi/n}$; hence m is a multiple of n, and it follows that $\mathbb{Q}_n$, generated by a power of $e^{2i\pi/m}$, is a subset of $\mathbb{Q}_m$. Finally, by double inclusion, $\mathbb{Q}_n = \mathbb{Q}_m$. Now, in order to clarify the relationship between n and m, we must consider the dimension of $\mathbb{Q}_n$ as a vector space on the rational field $\mathbb{Q}$.

It is $\varphi(n) = \dim[\mathbb{Q}_n/\mathbb{Q}]$, where φ is Euler's totient function[9], it stands that $n \mid m$ and $\varphi(n) = \varphi(m)$.

Since $\varphi(n) = n \times \prod\limits_{p|n \text{ and } p \text{ prime}} \left(1 - \dfrac{1}{p}\right)$, the only possibility is that $m = \begin{cases} n & \text{for } n \text{ even} \\ 2n & \text{for } n \text{ odd} \end{cases}$.

This proves that all eigenvalues of $\mathscr{U}$ are n^{th} or $2n^{th}$ roots of unity. Let us clarify the case of odd n: $e^{i\pi/n} = -\left(e^{2i\pi/n}\right)^{\frac{n+1}{2}}$ and hence we do indeed have $\mathbb{Q}_n = \mathbb{Q}_{2n}$. So we can rephrase what we just proved as Lemma 2.12: in the odd case, $\xi^n = \pm 1$.

---

[7] For instance $\mathbb{Q}_3 = \mathbb{Q}_6$, see exercises.

[8] It is perhaps not obvious that this group is finite, and indeed the group of elements of $\mathbb{Q}_2$ with magnitude one is not; essentially, this holds because for large m the dimension $\varphi(m)$ of the galoisian extension $\mathbb{Q}_m/\mathbb{Q}$ tends to infinity and thus exceeds $\varphi(n)$, dimension of $\mathbb{Q}_n/\mathbb{Q}$ (a more precise computation will be given in the main proof); hence roots of order m for large m cannot exist in $\mathbb{Q}_n$.

[9] This follows from the fact that the minimal polynomial of $e^{2i\pi/n}$ over $\mathbb{Q}$ is the cyclotomic polynomial Φ_n with degree $\varphi(n)$.

Remark 2.17. At this point, $\mathcal{U}$ could be constructed as a polynomial in the elementary circulating matrix $\mathcal{J}$ (as all other circulating matrixes) $\mathcal{U} = P(\mathcal{J})$, where P is the interpolating polynomial that sends the Fourier coefficients of $\mathcal{J}$, i.e. the $e^{2ik\pi/n}$, to the Fourier coefficients chosen for $\mathcal{U}$. Such a construction is easy and sometimes practical, using the basis of Lagrange polynomials associated with the $e^{2ik\pi/n}$, since P is a linear combination of these polynomials with coefficients that are precisely the Fourier coefficients of the desired u.

It is now time to prove Lemma 2.13: the possibilities of mapping the nth roots of 1 to m^{th} roots of 1 can be somewhat reduced by noticing that $\mathcal{U}$ is a rational polynomial[10] in $\mathcal{J}$, and such a polynomial is stable under all field automorphisms of $\mathbb{Q}_n$ if we use the following characterisation from Galois theory:

Lemma 2.18. *Any object (number, vector, polynomial, matrix) with coefficients in $\mathbb{Q}_n$ is rational iff it is invariant under all Galois automorphisms of the cyclotomic extension $\mathbb{Q}_n$ over $\mathbb{Q}$.*

We mention the structure of its Galois group without proof either.[11]

Lemma 2.19. *Any field (Galois) automorphism of the cyclotomic extension $\mathbb{Q}_n$ over $\mathbb{Q}$ is defined by $\Psi_k(e^{2i\pi/n}) = e^{2ik\pi/n}$ for some definite $k \in \mathbb{Z}_n^*$, the group of invertible elements of the ring $\mathbb{Z}_n$, e.g. for any integer k coprime with n.*

This is enough to define $\Psi_k(x)$ for any $x \in \mathbb{Q}_n$, since any element of $\mathbb{Q}_n$ can be written $x = \sum a_j e^{2ij\pi/n}$ with rational a_j's, and it follows that $\Psi_k(x) = \sum a_j e^{2ijk\pi/n}$. For instance when $n = 12$, there are exactly four different automorphisms Ψ_k, defined by the possible images of $e^{2i\pi/12} = e^{i\pi/6}$, namely $e^{ik\pi/6}$, $k \in \{1, 5, 7, 11\}$. Their group (the Galois group of the cyclotomic field) is isomorphic with the multiplicative group $\mathbb{Z}_{12}^* = \{1, 5, 7, 11\}$.

If Ψ_k is such an automorphism, notice that $\Psi_k(\xi) = \xi^k$ for any n^{th} root ξ of unity (with one exception: $\Psi_k(-1) = -1 \ \forall k \in \mathbb{Z}_n^*$). If n is odd and ξ is a $2n^{th}$ root but not a nth, then $-\xi$ is a nth root, and hence

$$\Psi_k(\xi) = -(-\xi)^k = \begin{cases} \xi_k & \text{for } k \text{ odd} \\ -\xi_k & \text{for } k \text{ even} \end{cases}.$$

For instance $\Psi_2(\xi) = -\xi^2$ for such ξ.

So from Lemma 2.18, we state that $\mathcal{U} \in \mathcal{M}_n(\mathbb{Q})$ iff $\mathcal{U}$ is invariant under all the $\Psi_k, k \in \mathbb{Z}_n^*$.

Now at last we can prove Lemma 2.13.

First case: n even.

Consider the eigenvector $X_j = (1, e^{2ij\pi/n}, e^{2i2j\pi/n}, \dots e^{2ij(n-1)\pi/n})^T$ for the eigenvalue ξ_j of $\mathcal{U}$ (for matrix J, the eigenvalue is of course $e^{2ij\pi/n}$). The ^{T} indicates

---

[10] The coefficients of this polynomial can be read on the first column of $\mathcal{U}$.

[11] These two results can be found in any textbook on Galois theory.

that we consider X_j as a column. We have $\Psi_k(X_j) = X_{jk}$ by direct computation. The case $j = 0$ is straightforward: since the eigenvector is real valued, so must be the eigenvalue, i.e. $\xi_0 = \pm 1$. We shall now set this case aside.

We assume that $\Psi_k(\mathcal{U}) = \mathcal{U}$ (i.e. that $\mathcal{U}$ is rational valued). Applying the Galois automorphism Ψ_k to the equation

$$\mathcal{U}X_j = \xi_j X_j \quad \text{yields} \quad \Psi_k(\mathcal{U})\Psi_k(X_j) = \mathcal{U}X_{kj} = \xi_{kj}X_{kj} = \Psi_k(\xi_j)\Psi_k(X_j) = \xi_j^k X_{kj}.$$

Hence

$$\Psi_k(\xi_j) = \xi_j^k = \xi_{jk} \quad (\sharp)$$

for all $j \neq 0$ and all $k \in \mathbb{Z}_n^*$.

Now for the reciprocal. Assume the above equation $(\sharp)$ between the eigenvalues. We choose one Galois automorphism, Ψ_k (for some k coprime with n). Let us apply $\Psi_k(\mathcal{U})$ to any eigenvector X_j of $\mathcal{U}$; notice that $X_j = \Psi_k(X_{k^{-1}j})$ where $k^{-1}j$ is computed modulo n. Hence

$$\Psi_k(\mathcal{U})X_j = \Psi_k(\mathcal{U}X_{k^{-1}j}) = \Psi_k(\xi_{k^{-1}j}X_{k^{-1}j}) = \Psi_k(\xi_{k^{-1}j})\Psi_k(X_{k^{-1}j})$$

$$= \xi_{k^{-1}j}^k X_{kk^{-1}j} \quad \text{because } \Psi_k \text{ raises any root of 1 to the } k^{th} \text{ power}$$

$$= \xi_j X_j \quad \text{by our assumption on the eigenvalues.}$$

We have proved that $\Psi_k(\mathcal{U})$ does the same thing as $\mathcal{U}$ on any eigenvector. But the eigenvectors of $\mathcal{U}$ constitute a basis, hence $\Psi_k(\mathcal{U}) = \mathcal{U}$ for all k coprime with n, i.e. $\mathcal{U}$ is rational valued.

Last case: n odd.

We still get the equation $\Psi_k(\xi_j) = \xi_{jk}$ if $\mathcal{U}$ is assumed to be invariant under Ψ_k. If ξ is a nth root of unity, the computation is identical.

If $\xi^{2n} = 1$ but $\xi^n \neq 1$, then $(-\xi)^n = 1$ and hence $\Psi_k(\xi) = -\Psi_k(-\xi) = -(-\xi)^k = -\xi^k$ for even k and $\Psi_k(\xi) = \xi^k$ for odd k. The computation above still yields $\xi_{jk} = \Psi_k(\xi_j) = \xi_j^k$ for odd k, and we have also the new case $\xi_{jk} = -\xi_j^k$ for even k.

Say $k = 2$, and $\xi_1 = \xi$ with $\xi_1^{2n} = 1 \neq \xi_1^n$; then $\xi_2 = -\xi^2, \xi_4 = -\xi^4, \ldots \xi_{2m} = -\xi^{2m}$. Number 2 has finite order in $\mathbb{Z}_n^*$, hence for some m, $\xi_{2m} = \xi_1$. We get an orbit of m eigenvalues which are all $2n^{th}$ roots of unity, e.g. $\mathcal{O} = \{\xi_1, \xi_2, \xi_4, \xi_8 \ldots\}$.

Say now that $k = 2^\nu k', k'$ odd and coprime with n. The formula $(\sharp)$ is then valid and yields $\xi_k = \xi_{2^\nu}^{k'}$. So ξ_k is determined when $\mathcal{O}$ is known. Notice that ξ_k will never be a nth root (because 2 and k' are coprime with n): either all the eigenvalues (with even index) are n^{th} roots, or none (except of course $\xi_0 = \pm 1$).

The reciprocal is similar to the even case:

- it is identical when the eigenvalues are of order n (at most); and
- if ξ_1 has order $2n$, then the values of ξ_k that we have obtained will satisfy the relations $\Psi_k(\mathcal{U})X_j = \xi_j X_j$ for all j, k §k coprime with n, so that $\Psi_k(\mathcal{U})$ is identical to $\mathcal{U}$, i.e. $\mathcal{U}$ is rational-valued.

This ends the proof of Lemma 2.13.

We can now use the cases expounded in Lemma 2.13 to prove Theorem 2.10.

The whole set of eigenvalues is determined if we know ξ_j for a subset of representatives j of all orbits under multiplication by elements of $\mathbb{Z}_n^*$ (so called *associated elements* in the ring $\mathbb{Z}_n$). We can specify the smallest representatives:

Lemma 2.20. *Any element $j \in$ the ring $\mathbb{Z}_n$ is associated with a divisor of n, i.e.* $\exists k \in \mathbb{Z}_n^*, kj = \gcd(n, j)$.

(We identify integers and classes modulo n here since the distinction is irrelevant).

Proof. This stems from the Bezout identity (in $\mathbb{Z}$): for some $k, \ell, kj + \ell n = \gcd(n, j)$. After division by $\gcd(n, j)$ we see that k and n are coprime. But modulo $n, kj = \gcd(n, j)$. ∎

Example 2.21. For $n = 15$ we have the orbits of equivalent elements

$$(0), (1, 2, 4, 7, 8, 11, 13, 14), (3, 6, 9, 12), (5, 10)$$

indexed by the divisors $1, 3, 5$ and of course 0. This classification will prove useful in Chapter 3.

So it is sufficient to specify ξ_j when j is any divisor of n. We will need a last lemma, interesting in its own right:[12]

Lemma 2.22. *The set of differences $\Delta_n = \mathbb{Z}_n^* - \mathbb{Z}_n^* = \{a - b, (a, b) \in (\mathbb{Z}_n^*)^2\}$ is $\mathbb{Z}_n$ when n is odd, $2\mathbb{Z}_n$ when n is even.*

Proof. It is straightforward for n prime, for n an odd prime power, and we notice that when $n = 2^m$ then $\mathbb{Z}_n^* =$ odd numbers, so that $\Delta_n =$ even numbers. The general case now stems from the Chinese remainder theorem, i.e. the Sylow decomposition: if $n = 2^d \ldots p^z \ldots$ is the prime decomposition of n then $\mathbb{Z}_n^* = (\mathbb{Z}/2^d\mathbb{Z})^* \times \ldots (\mathbb{Z}/p^z\mathbb{Z})^* \times \ldots$ and the result being true for the factors is true for the product. ∎

We now proceed to prove the theorem. Remember that $\xi_0 = \pm 1$.

- When n is even:
  In this case all eigenvalues are n^{th} roots of unity. Let j be any strict divisor of n.
  - When n/j is even, we can produce $k, k' \in \mathbb{Z}_n^*$ with $k' - k = \dfrac{n}{j} \in 2\mathbb{Z}$ from Lemma 2.22. Hence (noting that $k \equiv k' \mod n$)

$$\xi_j^{k + \frac{n}{j}} = \xi_j^{k'} = \xi_{jk'} = \xi_{jk} = \xi_j^k$$

  which proves that $\xi_j^{\frac{n}{j}} = 1$, i.e. ξ_j is a power of $e^{2ij\pi/n}$.

---

[12] Though elementary in nature, the result was previously unknown to the author and does not appear to be readily available in the literature.

- If n/j is odd (meaning that j contains the same power of 2 as n), then Lemma 2.22 only provides $k' - k = \dfrac{2n}{j}$, and the calculation yields $\xi_j^{2n/j} = 1$, i.e. ξ_j is a power of $e^{ij\pi/n}$, which ends the even case of the theorem.

- When n is odd:

  The case when ξ_j is a nth root is identical to the n even (first) case, as from the last Lemma 2.22, we can again produce two elements $k, k' \in \mathbb{Z}_n^*$ such that $k' - k = \dfrac{n}{j}$, and $\xi_j^{k'-k} = 1 = \xi_j^{n/j}$. So the spectral unit is a power of $e^{2i\pi j/n}$, i.e. a n/j^{th} root of unity ξ_j for each divisor j of n.

  Now assume that there is an eigenvalue ξ_j which is *not* a nth root. Then $-\xi_j$ is a nth root, and (as n/j is odd) a similar calculation yields for $k' - k = n/j$, with $k, k' \in \mathbb{Z}_n^*$,

  $$(-\xi_j)^{k'} = -\xi_{jk'} = -\xi_{jk} = (-\xi_j)^k = (-\xi_j)^{k'-\frac{n}{j}}.$$

  Hence $-\xi_j$ is again a nth root of unity, this is the second subcase.
  This ends the proof of the odd case of Theorem 2.10.

Theorem 2.11 follows from the possible independent values for each $\xi_j, j \mid n$: in general each ξ_j lies in a cyclic group with order n/j, while n/j runs over the list of divisors of n. The complicated situation is the case when $-\xi_j$ is also a n/j^{th} root, which explains the lcm in the formula (the group $\{\pm 1\} \times \mathbb{Z}_d$ is isomorphic with $\mathbb{Z}_{2d}$ whenever d is odd).

Example 2.23. This theorem enables us to find alternative spectral units between homometric pc-sets with some nil Fourier coefficients.

An example issued from music theory: consider two melodic minor scales $a = (1, 0, 1, 1, 0, 1, 0, 1, 0, 1, 0, 1), b = (1, 0, 1, 0, 1, 0, 1, 0, 1, 1, 0, 1)$. Their Fourier coefficients with indexes 2 and 10 are nil. Let us find a spectral unit u such that $a * u = b$, we have several possible choices for $\mathscr{F}_u(2)$:

- Using Rosenblatt's choice, we choose arbitrarily $\xi_2 = \mathscr{F}_u(2) = \mathscr{F}_u(10) = \xi_{10} = 1$ (the other Fourier coefficients are determined by $\mathscr{F}_u(k) = \mathscr{F}_b(k)/\mathscr{F}_a(k)$). This yields $u = (0, 0, 0, 0, 0, 0, 0, 0, 0, 1, 0, 0)$. Musically this means that A minor (melodic) is transposed from C minor by a minor third, a foreseeable result!

- We know from Lemma 2.13 that ξ_2 must be some power of $e^{i\pi/3}$, ξ_{10} being its conjugate or inverse. This yields no less than **five** other possible units, e.g.

$$u = \frac{1}{4}(1, 0, -1, -1, 0, 1, 1, 0, -1, 3, 0, 1) \quad \text{or}$$

$$u = \frac{1}{12}(1, 2, 1, -1, -2, -1, 1, 2, 1, 11, -2, -1) \quad \text{or}$$

$$u = \frac{1}{6}(2, 1, -1, -2, -1, 1, 2, 1, -1, 4, -1, 1) \quad \text{or}$$

$$u = \frac{1}{12}(1, -1, -2, -1, 1, 2, 1, -1, -2, 11, 1, 2) \quad \text{or}$$

$$u = \frac{1}{4}(1, 1, 0, -1, -1, 0, 1, 1, 0, 3, -1, 0).$$

In a way this can be interpreted as additional, hidden symmetries between those two musical scales.

Example 2.24. Let us elucidate this subgroup of units when $n = 12$. Let u be a spectral unit with finite order, and $\xi_0, \ldots \xi_{11}$ its Fourier coefficients, i.e. the eigenvalues of the associated matrix $\mathcal{U}$. From Theorem 2.10 above, the relation $\xi_j^k = \xi_{jk}$ is satisfied for all four values of $k = 1, 5, 7, 11$.

- There are no conditions on ξ_1 which is any 12th root ξ of unity; its value specifies $\xi_5 = \xi^5$ and similarly ξ_7, ξ_{11}.
- ξ_2 must be a power of $e^{2 \times 2i\pi/12} = e^{i\pi/3}$. This determines also $\xi_{10} = \overline{\xi}_2$.
- Similarly ξ_3 is a power of $i = e^{i\pi/2}$. We have $\xi_9 = \overline{\xi}_3$.
- Since 12/4 is odd (special case), ξ_4 must be a power of $e^{i\pi/3}$, just like ξ_2. Here also, we find that $\xi_8 = \overline{\xi}_4$.
- $\xi_0 = \pm 1$ and $\xi_6 = \xi_{-6} = \xi_6^{-1}$ is a $12/6^{th}$ root of 1, i.e. $\xi_6 = \pm 1$.

To conclude: ξ_1 is any 12th root of unity, while ξ_2, ξ_3, ξ_4 are limited to subgroups, ξ_0 and $\xi_6 = \pm 1$. The structure of the group is then $\mathbb{Z}_{12} \times (\mathbb{Z}_6)^2 \times \mathbb{Z}_4 \times (\mathbb{Z}_2)^2$, with 6,912 elements like $(0, 0, 1/3, -1/3, 0, 0, -2/3, -1/3, 0, 0, 1/3, -1/3)$ or $(7/12, -1/6, 1/12, 1/12, -1/6, -5/12, -5/12, -1/6, 1/12, 1/12, -1/6, -5/12)$. The complete list is available online as a text file:

http://canonsrythmiques.free.fr/allSpectralUnitsZ_12.txt.

It can be expanded from the following list of generators:

$$\left(\frac{5}{6}, -\frac{1}{6}, -\frac{1}{6}, -\frac{1}{6}, -\frac{1}{6}, -\frac{1}{6}, -\frac{1}{6}, -\frac{1}{6}, -\frac{1}{6}, -\frac{1}{6}, -\frac{1}{6}, -\frac{1}{6} \right)$$

$$\left(\frac{2}{3}, \frac{1}{6}, -\frac{1}{6}, -\frac{1}{6}, \frac{1}{6}, -\frac{1}{3}, \frac{1}{3}, -\frac{1}{6}, \frac{1}{6}, \frac{1}{6}, -\frac{1}{6}, \frac{1}{3} \right)$$

$$\left(\frac{11}{12}, -\frac{1}{6}, -\frac{1}{12}, \frac{1}{12}, \frac{1}{6}, \frac{1}{12}, -\frac{1}{12}, -\frac{1}{6}, -\frac{1}{12}, \frac{1}{12}, \frac{1}{6}, \frac{1}{12} \right)$$

$$\left(\frac{5}{6}, -\frac{1}{6}, \frac{1}{6}, \frac{1}{6}, -\frac{1}{6}, -\frac{1}{6}, \frac{1}{6}, \frac{1}{6}, -\frac{1}{6}, -\frac{1}{6}, \frac{1}{6}, \frac{1}{6} \right)$$

$$\left(\frac{11}{12}, -\frac{1}{12}, \frac{1}{6}, -\frac{1}{12}, \frac{1}{12}, \frac{1}{6}, -\frac{1}{12}, \frac{1}{12}, \frac{1}{6}, -\frac{1}{12}, \frac{1}{12}, \frac{1}{6} \right)$$

$$\left(\frac{5}{6}, \frac{1}{6}, -\frac{1}{6}, \frac{1}{6}, -\frac{1}{6}, \frac{1}{6}, -\frac{1}{6}, \frac{1}{6}, -\frac{1}{6}, \frac{1}{6}, -\frac{1}{6}, \frac{1}{6} \right);$$

one may notice that the first one is the opposite of the complement operator, cf. Proposition 2.7.

2.1.4 Orbits for homometric sets

We have seen that the action of the torus of spectral units describes the most general orbits of homometric classes in the vector space $\mathbb{C}^n$, but fails to elicit the distributions in this space which are actual pc-sets, i.e. distributions with values 0 or 1.[13] Actually there is a deep result behind this failure:

Theorem 2.25. *Let $n \in \mathbb{N}$ with $n \geqslant 2$. If $n = 8$, $n = 10$ or $n \geqslant 12$, then for every field K and for every subgroup H of the linear group $GL_n(K)$ such that the natural group action of H on $\mathscr{P}(\mathbb{Z}_n)$ identified with $\{0,1\}^n$ is well-defined, the orbits of this group action are not identical with the equivalence classes of the Z-relation.*

This stunning result discovered by John Mandereau [64] needs translation: it means that there is no 'reasonable' group action (that would induce some action on the pcs themselves) whose orbits are the homometric classes.[14]

Of course, it is possible to study the symmetries of *one* class of isometric pc-sets as subgroups of the group of permutations of k-subsets. Such symmetry groups depend on the class and usually include (or coincide with) T/I. The other cases are intriguing: for instance the group of the homometry class of $\{0,1,4,6\}$ in $\mathbb{Z}_{12}$ is isomorphic with the 48-element affine group modulo 12.[15] The drawback of this topdown approach is that the homometry class has to be computed before the symmetry group. On the other hand, elucidating the relationships between the elements of an homometry class is extremely useful for composers: for instance, the aforementioned class is composed of one orbit under the affine group, two orbits under T/I ($\{0,1,4,6\}$ and $\{0,1,3,7\}$) and four under T (adding $\{0,2,5,6\}$ and $\{0,4,6,7\}$, see Fig. 8.13). More about the computations of these groups can be found in [41], hinting at some compositional applications by Tom Johnson. A rich example uses paths between the 108 homometric sets with size 5 in $\mathbb{Z}_{12}$, computed by Franck Jedrzejewski and drawn by Johnson in Fig. 2.4, each line corresponding with one of three generators a, b, c of the symmetry group.

2.2 Extensions and generalisations

2.2.1 Hexachordal theorems

We have stated the original hexachord theorem in modern terms:

Theorem 2.26 (Babbitt's hexachord theorem).
   Any hexachord in $\mathbb{Z}_{12}$ is homometric with its complement.

The proof can be easily adapted to a more general statement:

Theorem 2.27. *The intervallic contents of a subset of $\mathbb{Z}_n$ and of its complement differ by a constant distribution, whose value is the difference between the cardinality of the set and of its complement:*

---

[13] It is possible to get down to $\mathbb{R}^n$ but even the difficult Theorem 2.10 does not completely elucidate homometry in $\mathbb{Q}^n$, leaving aside infinite orbits.

[14] In this light one may remember that moving from a major to a minor triad was a 'local' transformation, depending on both triads.

[15] Actually it *IS* the affine group itself, permuting the interval vector without changing it since these are all-interval sets.

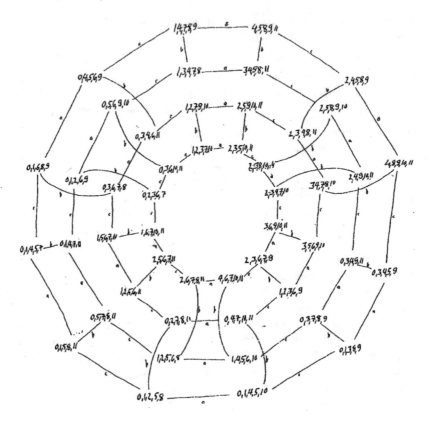

Fig. 2.4. Paths between homometric 5-sets drawn by T. Johnson

$$IC(A) - IC(\mathbb{Z}_n \setminus A) = (\#(A) - \#(\mathbb{Z}_n \setminus A)).\mathbf{1} = Constant$$

For instance, the intervallic contents of Tristan's chord $\{3,5,8,11\}$ and its complement are $(4,0,1,2,1,2,0,2,1,2,1,0)$ and $(8,4,5,6,5,6,4,6,5,6,5,4)$; the difference between those two vectors is constant and equal to 4, the cardinality difference. If one is sensitive to the ratios between the interval counts, then the intervallic distribution is clearer on the smaller pc-set. If however one perceives the variation of this interval histogram around a mean value, then perhaps the larger complement set yields a neater intervallic distribution: the theorem says that the two ICs are equal (up to a constant) but the *contrast* differs. In optics this result is actually known as Babinet's theorem:

> The diffraction pattern from an opaque body is identical to that from a hole of the same size and shape except for the overall forward beam intensity.

The difference in contrast means that, for instance, one can estimate the breadth of a hair by carving it out of an opaque sheet and diffracting (ordinary) light with this,

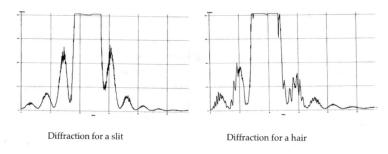

Diffraction for a slit   Diffraction for a hair

Fig. 2.5. Two diffraction graphs for a slit and its complement

or estimate the size of red blood cells by comparing the diffraction picture with one obtained from calibrated small holes, see [20] and Fig. 2.5.[16]

This theorem can be further extended to a large class of groups (including mainly compact groups), see [2]. The proof using Fourier transform is still valid for all finite abelian groups and even compact abelian groups (such as the torus $\mathbb{T}_n$), but we will not spell it out here since there is a more general one. The essential point is that the probability of occurrence of an 'interval' g (i.e. the size of $\mathscr{I}_g = \{(a,b) \in G \times G, b = g.a\}$), can still be measured by integral calculus thanks to the existence of a Haar measure.[17]

A nice example in a torus is the following, borrowed from the above paper:

Example 2.28. Musical scales can be modelised as elements of a torus, each note being a point on the continuous unit circle S^1 (see Section 5.1). Say we define the set ITS of 'in-tune' scales as major scales whose maximal deviation from a reference well-tempered major scale does not exceed 10 cents, e.g. the 'in-tune' D major scales would be in $[190,210] \times [390,410] \times [590,610] \times [690,710] \times [890,910] \times [1090,1110] \times [90,110]$ where each pc is given in cents. So ITS is a subset of the torus $\mathbb{T}^7 = (\mathbb{R}/1200\,\mathbb{Z})^7$, with measure $(20/1200)^7$ of the whole torus, and the complement OTS (out-of-tune' scales) has the same interval content, up to a constant.

The simplest generalisation is to finite abelian groups, which are products of cyclic groups (i.e. discrete torii). Such a group can model for instance:

1. The decomposition $\mathbb{Z}_{12} = \mathbb{Z}_3 \times \mathbb{Z}_4$, so-called torus of thirds: the hexachord theorem (or the notion of homometry in general) can be factored down to this expression of pcs as pairs.
2. Pairs (or p−uples) of pcs lie in $\mathbb{Z}_{12} \times \mathbb{Z}_{12}$ (or a larger power), wherein the general hexachord theorems apply.

---

[16] Obtained by one of my students, Domenech Vianney, in 2015.

[17] This means that there is a way to measure a subset's 'size' which is invariant under translation.

2.2.2 Phase retrieval even for some singular cases

As discussed above, knowledge of $\mathrm{IFunc}(A,B)$ and B enables us to retrieve A, except when $\mathscr{F}_B$ vanishes because $\mathscr{F}_A$ is then indeterminate. It is still possible though to retrieve A, solving Lewin's problem, when $\mathscr{F}_B$ vanishes for a *single* coefficient[18] and when A is known to be a genuine set, i.e. a distribution with only 0's and 1's. This additional information compensates for the missing one. It is perhaps best to describe the somewhat involved process by way of an example:

Example 2.29. The melodic minor scale $B = \{0,2,3,5,7,9,11\}$ is one of Lewin's special cases: $\mathscr{F}_B(2)(=\mathscr{F}_B(10)) = 0$.

Assume $\mathrm{IFunc}(A,B) = (2,2,2,1,2,3,0,3,1,2,2,1)$ where A remains to be found. Adding the number of intervals, i.e. the elements of $\mathrm{IFunc}(A,B)$, one gets 21, meaning that A has three elements to B's seven. Now compute all available values of $\mathscr{F}_A$, i.e. all except $\mathscr{F}_A(2), \mathscr{F}_A(10)$, dividing the coefficients of the DFT of $\mathrm{IFunc}(A,B)$ by the conjugates of those of B. One gets (rounding to the third digit for legibility)

$$\mathscr{F}_A = (3., -0.366 - 0.366i, \boxed{\mathbf{X}}, 2+i, -1.732i, 1.366 + 1.366i, 1, \text{ [and conjugates]}).$$

The secret weapon at this juncture is Parseval-Plancherel's formula:

$$\sum |\mathscr{F}_A(k)|^2 = n\#A$$

in the case of a pc-set. This provides the *magnitude* of the missing Fourier coefficient $\mathbf{X}$, the phase φ being still unknown: let $\mathbf{X} = \mathscr{F}_A(2) = \overline{\mathscr{F}_A(10)} = re^{i\varphi}$, then the difference between the sum of all known $|\mathscr{F}_A(k)|^2$ (here equal to 34) and $n\#A = 3 \times 12 = 36$, is equal to $2r^2$. Hence $r = 1$. Plugging back in this value, we are now down to

$$\mathscr{F}_A = (3., -0.366 - 0.366i, e^{i\varphi}, 2+i, -1.732i, 1.366 + 1.366i, 1, \text{ and their conjugates})$$

By inverse Fourier transform, we get (I only quote the first values)

$$\mathbf{1}_A = \left(\frac{\cos(\varphi)}{6} + 0.833, \frac{1}{6}\sin\left(\frac{\pi}{6} - \varphi\right) - 0.083, 0.083 - \frac{1}{6}\sin\left(\varphi + \frac{\pi}{6}\right), \ldots \right)$$

Now the only way $\dfrac{\cos(\varphi)}{6} + 0.833333$ can be equal to 0 or 1 is to have $\cos\varphi = 1$, i.e. $\varphi = \pm\dfrac{2\pi}{3}$. The value of φ could be found equally easily from any other coefficient, e.g. $\frac{1}{6}\sin\left(\frac{\pi}{6} - \varphi\right) = 0.08333$ would yield the same solution (in other cases, it might be necessary to examine several equations in order to dispel possible ambiguities – or perhaps find multiple solutions).

Plugging this value of φ in $\mathbf{1}_A$ finally yields (up to rounding errors)

$$\mathbf{1}_A = (1,0,0,0,1,0,0,1,0,0,0,0),$$

i.e. $A = \{0,4,7\}$ which was indeed the pc-set that served to compute $\mathrm{IFunc}(A,B)$ in the first place.

---

[18] And of course its conjugate.

Of course, in this particular case it might be quicker to proceed by trial and error, but the method is general. To sum up the algorithm, one follows these steps:

1. Compute the cardinality of A: it is the sum of the elements of $\text{IFunc}(A,B)$ divided by $\#B$.
2. Compute $\mathscr{F}_A = \dfrac{\mathscr{F}(\text{IFunc}(A,B))}{\mathscr{F}_B}$, with two coefficients still indeterminate.
3. Compute the sum of the squared magnitudes of the $n-2$ known coefficients in the last step; subtract the result from $n\#A$ to get $2r^2$ and hence r, the magnitude of the missing coefficient.
4. Compute the inverse Fourier transform of $\mathscr{F}_A$ as a function of the missing coefficient $re^{i\varphi}$, where only φ remains unknown.
5. Taking into account that all the values computed in the last step must be 0's or 1's, determine φ; complete the computation of $\mathbf{1}_A$.

To some extent, this algorithm could be used even when A is a multiset.

For practical purposes, I will remind the reader of the matricial formalism mentioned in 1.2.3. In [13], we used linear programming to good effect for solving equations like $s * \mathbf{1}_A = \mathbf{1}_B$ (which corresponds to finding a linear combination of translates of A equal to B) and the same procedure could be used for solving $\mathbf{1}_A * \mathbf{1}_{-B} = \text{IFunc}(A,B)$ in A, which is the problem at hand. But though the algorithm seems to work well, it is not formally proved yet that it always provides a solution. For one thing, there may well be multiple solutions (that the algorithm may reach by varying the starting point), e.g. for $B = \{0,2,4,6,8,10\} \subset \mathbb{Z}_{12}$, $\text{IFunc}(A,B)$ does not change when A is replaced by $A+2$. See the reference above or 3.3.3 for a description of this method, which bypasses Fourier transform altogether.

2.2.3 Higher order homometry

IFunc counts intervals, which are pairs of elements. There is no law against counting triplets, quadruplets, and so on. It is necessary to be precise about what is a different 'occurrence' of a given triplet. We borrow again some definitions and results from [64], with some modifications.

Let us begin with counting triplets (i.e. 3-subsets of a pc-set) up to translation: if we are looking for copies of $(0,a,b)$, their number in $A \subset \mathbb{Z}_n$ is equal to

$$\mathbf{tv}(a,b) = \sum_{t \in \mathbb{Z}_n} \mathbf{1}_A(t)\mathbf{1}_A(t+a)\mathbf{1}_A(t+b).$$

We redo from scratch the computation of the Fourier transform, here in two variables:[19]

---

[19] All sums are taken over the whole $\mathbb{Z}_n$.

$$\widehat{tv}(\omega,v) = \sum_t \sum_a \sum_b \mathbf{1}_A(t)\mathbf{1}_A(t+a)\mathbf{1}_A(t+b)e^{-2i\pi(\omega a+vb)/n}$$

$$= \sum_t \mathbf{1}_A(t)e^{2i\pi(\omega t+vt)/n}\sum_a \mathbf{1}_A(t+a)e^{-2i\pi\omega(t+a)/n}\sum_b \mathbf{1}_A(t+b)e^{-2i\pi v(t+b)/n}$$

$$= \sum_t \mathbf{1}_A(t)e^{-2i\pi(-\omega-v)t/n}\sum_x \mathbf{1}_A(x)e^{-2i\pi\omega x/n}\sum_y \mathbf{1}_A(y)e^{-2i\pi vy/n}$$

$$= \mathscr{F}_A(-\omega-v)\mathscr{F}_A(\omega)\mathscr{F}_A(v).$$

Hence

Proposition 2.30. *The triplet histograms of pc-sets A and B are equal iff for all $\omega, v \in \mathbb{Z}_n$*

$$\mathscr{F}_A(-\omega-v)\mathscr{F}_A(\omega)\mathscr{F}_A(v) = \mathscr{F}_B(-\omega-v)\mathscr{F}_B(\omega)\mathscr{F}_B(v).$$

Generalizing to k-uplets, we will say that A, B are **k-homometric,** *i.e. contain the same number of translates of any k-subset, or more generally that two distributions E and F are k-homometric, iff*

$$\widehat{E}(\omega_1)\widehat{E}(\omega_2)\cdots\widehat{E}(\omega_{k-1})\widehat{E}(-\omega_1-\ldots-\omega_{k-1}) =$$
$$= \widehat{F}(\omega_1)\widehat{F}(\omega_2)\cdots\widehat{F}(\omega_{k-1})\widehat{F}(-\omega_1-\ldots-\omega_{k-1})$$

for every $(\omega_1,\ldots,\omega_{k-1}) \in \mathbb{Z}_n^{k-1}$.

It is easily seen from this formula that

1. k-homometry implies $(k-1)$-homometry,[20] and
2. 2-homometry is usual homometry[21]:

$$|\mathscr{F}_A(\omega)|^2 = \mathscr{F}_A(\omega)\mathscr{F}_A(-\omega) = \mathscr{F}_B(\omega)\mathscr{F}_B(-\omega) = |\mathscr{F}_B(\omega)|^2.$$

The study of phase retrieval (find all distributions k-homometric with a given E) is hence very difficult when the Fourier transform vanishes. When it does not, there is a strong result:

Theorem 2.31. *When E is non negative and if $\widehat{E}$ never vanishes on $\mathbb{Z}_n$, any distribution 3-homometric with E must be a translate of E.*

The proof illustrates the strength and relevancy of the DFT.

Proof. Assume E and F are 3-homometric, $\widehat{E}$ and $\widehat{F}$ never vanishing. Let us denote by ξ the ratio of the DFTs, $\xi(t) = \widehat{E}(t)/\widehat{F}(t)$. Then from Proposition 2.30 we get that

$$\xi(\omega+v) = \xi(\omega)\xi(v) \quad \forall \omega, v \in \mathbb{Z}_n;$$

---

[20] At least when $\widehat{F}(0) \neq 0$.

[21] By now surely nobody will presume to call it 'simple homometry'.

meaning that ξ is a group morphism from $(\mathbb{Z}_n, +)$ into $(\mathbb{C}^*, \times)$, a.k.a. a *character*. The characters of $\mathbb{Z}_n$ are well-known: there is[22] an integer k such that $\xi : t \mapsto e^{-2i\pi kt/n}$. But this means

$$\forall t \in \mathbb{Z}_n \; \widehat{E}(t) = \widehat{F}(t) \times e^{-2i\pi kt/n}.$$

By inverse Fourier transform (or by reversing Proposition 1.16) this means that $E = F + k$.

Here is an example of non-trivial 3-homometry in $\mathbb{Z}_{32}$:

$$A = \{0, 7, 8, 9, 12, 15, 17, 18, 19, 20, 21, 22, 26, 27, 29, 30\},$$
$$B = \{0, 1, 8, 9, 10, 12, 13, 15, 18, 19, 20, 21, 22, 23, 27, 30\}.$$

These sets are 3-homometric – for instance the pattern (0, 10, 20) appears seven times in both – but not translates, cf. Fig. 2.6 (hence their DFT must vanish; indeed all Fourier coefficients with even index are nil).

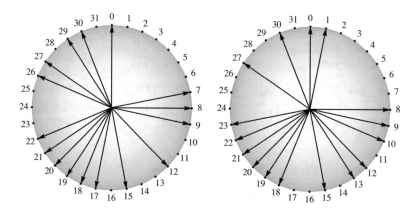

Fig. 2.6. Two 3-homometric subsets

This result narrows the import of the notion of k-homometry of pc-sets: in most cases, this notion is nothing new since it reduces to equivalence under translation.[23] This is probably why the literature usually addresses a broader form of homometry. Indeed a problem appears for $k \geqslant 3$ which did not make sense for $k = 2$, i.e. when

---

[22] Such a morphism is determined by the image of $1 \in \mathbb{Z}_n$ since one element generates the whole group. We used a stronger form of this result during the proof of Theorem 2.10.

[23] The result is still true even in several cases with vanishing DFT, when n has few factors, though the proof gets really difficult (see [64], Section 4). We will see though in the next chapter that distributions with nil Fourier coefficients play vital roles in some areas of music theory, so perhaps this area deserves further research. For instance, both subsets given in the last example tile (trivially) $\mathbb{Z}_{32}$.

counting intervals: clearly whenever an interval appeared, so did its inverse. But for triplets or larger subsets, the inversion is usually a distinct form. Hence the following, taken again from [64], Section 4:

Definition 2.32. *Let H be a subgroup of the group of permutations of $\mathbb{Z}_n$, $S(\mathbb{Z}_n)$. Let us define a H-copy of a set $S \subset \mathbb{Z}_n$ as any set of the form $h(S)$, with $h \in H$. Their set, the orbit of S under the action of H, will be denoted by $[S]_H$.*

The two most interesting cases are $H = T$, the cyclic group of transpositions, and $H = T/I$, the dihedral group of transpositions and inversions, though other groups, like the affine group, might be of interest for composers.

Definition 2.33. *Let $A \subset \mathbb{Z}_n$; we call k-vector of A the map*

$$S \mapsto \mathbf{mv}^k(A)_S = \#\{S' \in [S]_{T/I}, S' \subset A\}.$$

For of any k-set S, it tallies the number of its T/I-copies embedded in A.

Example 2.34. The set $A = \{0,1,3,4,7\}$ has essentially only six non-zero entries in its 3-vector:

$$\mathbf{mv}^3(A)_{\{0,1,3\}} = 2 \qquad\qquad \mathbf{mv}^3(A)_{\{0,1,4\}} = 3$$
$$\mathbf{mv}^3(A)_{\{0,1,6\}} = 1 \qquad\qquad \mathbf{mv}^3(A)_{\{0,2,6\}} = 1$$
$$\mathbf{mv}^3(A)_{\{0,3,6\}} = 1 \qquad\qquad \mathbf{mv}^3(A)_{\{0,3,7\}} = 2$$

Indeed, $\mathbf{mv}^3(A)_{\{0,1,3\}} = 2$ since there are two T/I-copies of $\{0,1,3\}$ embedded in A (they are $\{0,1,3\}$ and $\{1,3,4\}$); $\mathbf{mv}^3(A)_{\{0,1,4\}} = 3$ since there are three T/I-copies of $\{0,1,4\}$ embedded in A (they are $\{0,1,4\}$, $\{0,3,4\}$ and $\{3,4,7\}$); and so on.

This is more general than what we have done with k-homometry.

Definition 2.35. *Sets $A_1, \ldots, A_s$ are k-Homometric (with a capital 'H') iff $\mathbf{mv}^k(A_1)_S = \mathbf{mv}^k(A_2)_S = \ldots = \mathbf{mv}^k(A_s)_S$ for all $S \subset \mathbb{Z}_n$, $\#S = k$.*

Example 2.36. Let us consider, in $\mathbb{Z}_{18}$, the two sets $A = \{0,1,2,3,5,6,7,9,13\}$ and $B = \{0,1,4,5,6,7,8,10,12\}$. They are not related by translation/inversion, but $\mathbf{mv}^3(A)_S = \mathbf{mv}^3(B)_S$ for all $3-$subsets S. For instance the set $S = \{0,1,9\}$ appears once in A and once, inverted, in B (see Fig. 2.7).

Their Fourier transform never vanishes, which shows that Theorem 2.31 works with general homometry (by translation) but not with Homometry (by translation/inversion).

The search for non-trivial k-Homometry is a formidable computational problem, but an example for $k = 4$ was found in 2011 by Daniele Ghisi.

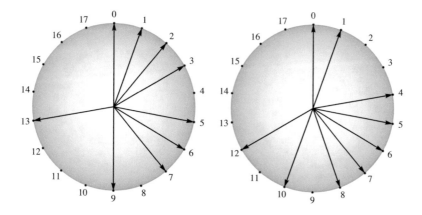

Fig. 2.7. Two non-trivially 3-Homometric subsets of $\mathbb{Z}_{18}$

Exercises

Exercise 2.37. Choose one hexachord, compute its intervallic distribution and that of its complement. Are these two hexachords T/I related?

Exercise 2.38. Compose a melody with four notes in $\{0,1,4,6\}$ in one of its translated forms (say B C E♭ F) spelling eleven distinct intervals. Superimpose another melody with the same intervals, but taken in a homometric pc-set, say $\{0,1,3,7\}$.

Exercise 2.39. Find non-trivially homometric pentachords (two classes). Are they affinely related?

Exercise 2.40. Prove Theorem 2.2 for non-singular distributions (i.e. their DFT never vanishes).

Exercise 2.41. Prove Proposition 2.7 by computing the eigenvalues and eigenspaces and the convolution product with an arbitrary characteristic function.

Exercise 2.42. Compute some non-obvious cubic roots of the circulating matrix of the minor third transposition $mt = (0,0,0,1,0,0,0,0,0,0,0,0) = j^3$ in $\mathscr{C}_n(\mathbb{C})$ (hint: use the matrix formalism and eigenvalues). Which of the solutions belong to $\mathscr{C}_n(\mathbb{R}), \mathscr{C}_n(\mathbb{Q})$?

Exercise 2.43. Cyclotomic fields: find a linear basis of $\mathbb{Q}_3$ over field $\mathbb{Q}$. Same thing with $\mathbb{Q}_6$, checking that $\mathbb{Q}_6 = \mathbb{Q}_3$.

3

Nil Fourier Coefficients and Tilings

Summary. Originally, vanishing Fourier coefficients appeared as an obstruction: they impede phase retrieval and prevent, for instance, the solution of Lewin's problem (find A knowing B and $\mathrm{IFunc}(A,B)$). But recent research and problems shed a more positive light: for instance the set $Z(A)$ of indexes k such that $a_k = \widehat{\mathbf{1}_A}(k) = 0$, a highly organised subset of $\mathbb{Z}_n$, is now the fashionable introduction to a definition of tilings. The theory of tilings is a crossroad of geometry, algebra, combinatorics, topology; and one of those privileged domains where musical ideas enable us to make some headway in non-trivial mathematics. Here the notion of Vuza canon together with transformational techniques (often introduced by composers) allowed some progress on difficult conjectures. More generally, tiling situations provide rich compositional material as we will see later in this book, cf. Section 4.3.3. In that respect, I included in this chapter Section 3.3 on algorithms (for practical purposes, though there are some interesting theoretical implications in there too.

We will need some additional algebraic material on polynomials, which is introduced in the preliminary section. A few more technical results of Galois theory are recalled and admitted without proof.

Cyclotomic polynomials

We will require the notion of cyclotomic polynomial. The etymology is telling: much of our work relates to 'splitting the circle', and this notion is the most powerful tool to do it.

Lemma 3.1. *Let* $\Phi_m(X) = \prod(X - \xi)$ *where* ξ *runs over the set of roots of unity with order exactly* m, *i.e.* $\xi^m = 1$ *but* $\xi^p \neq 1$ *for* $0 < p < m$. *In other words,*

$$\Phi_m(X) = \prod_{k \in \mathbb{Z}_m^*} (X - e^{2i\pi k/m}).$$

Then $\Phi_m \in \mathbb{Z}[X]$ *(it has integer coefficients) and* Φ_m *is irreducible in the ring* $\mathbb{Q}[X]$ *(or* $\mathbb{Z}[X]$*): any divisor of* Φ_m *is a constant or* Φ_m *itself.*

Proof. The non-obvious point is the irreducibility in $\mathbb{Q}[X]$, we refer the curious reader to textbooks or the Internet. The integral character of the coefficients derives

© Springer International Publishing Switzerland 2016
E. Amiot, *Music Through Fourier Space*, Computational Music Science,
DOI 10.1007/978-3-319-45581-5_3

from the following formula, each polynomial being monic. It is also an effective way of computing these polynomials by Euclidean division:

$$X^n - 1 = \prod_{d|n} \Phi_d(X) \tag{3.1}$$

For instance for $n = p$ prime, we get $\Phi_p(X) = \dfrac{X^p - 1}{X - 1} = 1 + X + \ldots X^{p-1}.$

The meaning of this is that any rational polynomial which vanishes in some root of unity must be divisible by Φ_m, i.e. it also features all other roots with the same order. Actually this is one way to prove the irreducibility, using the Galois automorphisms of the cyclotomic field which permutes roots with the same order so that any polynomial featuring the factor $(X - \xi)$ in $\mathbb{C}[X]$ also features $(X - \xi')$ if ξ' has the same order.

By induction one derives the following from formula 3.1:

Proposition 3.2. $\Phi_n(1)$ *is equal to* $\begin{cases} p & \textit{if } n \textit{ is a prime power } p^\alpha \\ 1 & \textit{else} \end{cases}.$

3.1 The Fourier nil set of a subset of $\mathbb{Z}_n$

3.1.1 The original caveat

It is now clear that when Lewin wrote his first paper [62] wherein he considered the question of identifying A from the knowledge of another pc-set B and $\mathrm{IFunc}(A, B)$, he had in mind the formula

$$1_A * 1_{-B} = \mathrm{IFunc}(A, B) \iff \mathscr{F}_A \times \overline{\mathscr{F}_B} = \widehat{\mathrm{IFunc}}(A, B).$$

However he could only allude to Fourier transform (and even that earned him outraged reactions from readers of the *Journal of Music Theory*). So perhaps he was right in stating the condition that $\mathscr{F}_B$ vanished in less mathematical terms. However, 'Lewin's conditions' are far from convenient. Let us enumerate these cases[1] which prevent[2] recuperation of one pc-set from its intervallic relationship with another:

1. **the whole-tone scale property**
   A chord has this property if it "has the same number of notes in one whole-tone set, as it has in the other [whole-tone set]."
2. **the diminished-seventh chord property**
   A chord has this property if it "has the same number of notes in common with each of the three diminished-seventh chord sets."
3. **the augmented triad property**
   A chord has this property if, "for any augmented-triad set A, [it] has the same number of notes in common with $T_6(A)$,[3] as it has in common with A."

---

[1] We use a more synthetic presentation [63] than the original one [62] which is frankly unreadable.

[2] See however the new method in Section 2.2.2 above.

[3] As usual in music theory, $T_k(A) = A + k$ denotes the transposition by k semitones.

4. **the tritone property**

   A chord has this property if "for any (0167)-set K, [it] has the same number of notes in common with $T_3(K)$, as it has in common with K." This is equivalent to keeping the difference of notes between the intersection with a tritone $\mathscr{T}$ and its translate $T_3(\mathscr{T})$ a constant, hence the original name.

5. **the exceptional property**

   A chord has this property if it "can be expressed as a disjoint union of tritone sets and/or augmented-triad sets" (the original definition enumerated no less than 10 sub-cases).

The least one can say about these properties (especially the last two) is that they are not exactly straightforward, especially when compared to the concise '$\mathscr{F}_A(k) = 0$' (respectively for $k = 6, 4, 3, 2, 1$ as we will develop below). More precisely, they originate in the nullity of several specific Fourier coefficients, respectively (at least)

1. The 6th for the whole-tone property;
2. The 4th and 8th for the diminished-seventh property;
3. The 3rd and 9th for the augmented triad property;
4. The 2nd and 10th for the "tritone" property;
5. The $1^{st}, 5^{th}, 7^{th}$ and 11th for the exceptional property.

See below the discussion around Theorem 3.11 for an explanation of this multiplication of nil Fourier coefficients. In Fig. 3.1, one can see an example for each situation, following the order in which they are enumerated in the text.

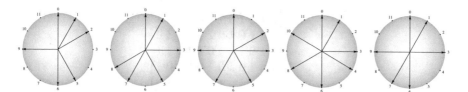

Fig. 3.1. The five special cases enumerated by Lewin

In his dissertation, Ian Quinn introduced a wonderfully telling implementation of these conditions, in terms of 'balances' (the word here is taken in the non-musical meaning of [weighing] 'scale', this word being admittedly misleading in the context). For instance, the third one is expressed by the balance of four pans, each containing the intersection of A with one of the four augmented triads, see Fig. 3.2. Though the expression of the five conditions with Quinn's balances has an aesthetic charm of its own, it is still cumbersome to check whether a given pc-set will fail one of them. We can provide a more synthetic characterisation of the 'bad cases' of Lewin's problem:

Theorem 3.3. *A distribution s has at least one nil Fourier coefficient iff the associated circulating matrix $\mathscr{S}$ is singular, which can be checked for instance with its determinant (or rank).*

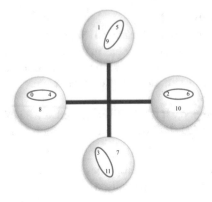

Fig. 3.2. Condition $\mathscr{F}_A(3) = 0$ is checked by pc-set $\{0,2,3,4,5,6,9,11\}$.

Example 3.4. One can check whether the melodic A minor $\{0,2,4,6,8,9,11\}$ is a 'bad case' by computing the following determinant, which is straightforward for most pocket calculators and does not involve the complex numbers and exponentials featured in the definition of the DFT:

$$\det(\mathscr{S}) = \begin{vmatrix} 1 & 1 & 0 & 1 & 1 & 0 & 1 & 0 & 1 & 0 & 1 & 0 \\ 0 & 1 & 1 & 0 & 1 & 1 & 0 & 1 & 0 & 1 & 0 & 1 \\ 1 & 0 & 1 & 1 & 0 & 1 & 1 & 0 & 1 & 0 & 1 & 0 \\ 0 & 1 & 0 & 1 & 1 & 0 & 1 & 1 & 0 & 1 & 0 & 1 \\ 1 & 0 & 1 & 0 & 1 & 1 & 0 & 1 & 1 & 0 & 1 & 0 \\ 0 & 1 & 0 & 1 & 0 & 1 & 1 & 0 & 1 & 1 & 0 & 1 \\ 1 & 0 & 1 & 0 & 1 & 0 & 1 & 1 & 0 & 1 & 1 & 0 \\ 0 & 1 & 0 & 1 & 0 & 1 & 0 & 1 & 1 & 0 & 1 & 1 \\ 1 & 0 & 1 & 0 & 1 & 0 & 1 & 0 & 1 & 1 & 0 & 1 \\ 1 & 1 & 0 & 1 & 0 & 1 & 0 & 1 & 0 & 1 & 1 & 0 \\ 0 & 1 & 1 & 0 & 1 & 0 & 1 & 0 & 1 & 0 & 1 & 1 \\ 1 & 0 & 1 & 1 & 0 & 1 & 0 & 1 & 0 & 1 & 0 & 1 \end{vmatrix} = 0$$

Remark 3.5. Another way to check that the matrix is singular consists of noticing that the sums of columns $1,6,7,12$ is the same as that of columns $3,4,9,10$, namely $(3,2,2,3,2,2,3,2,2,3,2,2)^T$.

In our opinion it is high time that a spade be called a spade, and 'Lewin's special cases' should be computed in the way they were discovered, i.e. by checking the nullity of Fourier coefficients.

Usually a clock diagram of the *multiset* $(kA)_{mult}$ (all multiples of elements of A, times k mod n, counted with their multiplicities) will enable one to see at a glance whether $\mathscr{F}_A(k) = 0$. In Fig. 3.3 one can see the diagrams for $\mathscr{F}_A(1)$ and $\mathscr{F}_A(2)$ where A is the melodic minor above. For the first coefficient, the clock represents just A and one cancels out 0-6 and 2-8; the remainder 4-9-11 obviously does not sum to 0. On the next clock, $(2A)_{mult} = \{0,4,8,0,4,6,10\}$ is a multiset with 0-4 redoubled. Gathering 0-4 together with 8 as a subset with sum 0 leaves 0-4-6-10 which also sums to nil. All cases of nil coefficients for $n = 12$ are similarly reducible to obvious cases (see Conjecture 3.16 and Fig. 3.5 below though), the 'special case' being actually the simplest, since no multiplication of A into a multiset is necessary.

A complete table of the 134 pc-sets classes (up to transposition) with some nil Fourier coefficient is provided on Table 8.2.[4]

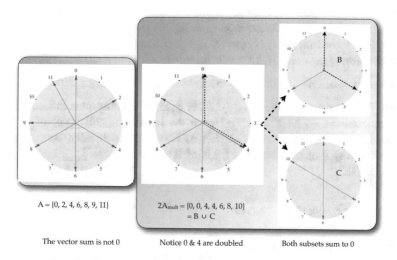

$A = \{0, 2, 4, 6, 8, 9, 11\}$ $2A_{mult} = \{0, 0, 4, 4, 6, 8, 10\}$
$= B \cup C$

The vector sum is not 0   Notice 0 & 4 are doubled   Both subsets sum to 0

Fig. 3.3. Checking nullity of some Fourier coefficients

3.1.2 Singular circulating matrixes

According to Theorem 3.3, the vanishing of some Fourier coefficients can be checked by computing a determinant. We introduce the corresponding matricial vocabulary for convenience:

Definition 3.6. *A distribution $s \in K^n$ is* singular $\iff \det \mathscr{S} = 0$, *i.e. when at least one of its Fourier coefficients is nil ($\mathscr{S}$ is the circulating matrix associated with s). Otherwise it is* invertible.

From the characterisation of singular matrixes by the linear dependency of their columns we get the useful

Proposition 3.7. *A subset A of $\mathbb{Z}_n$ is singular iff the subset is a linear combination of its translates $A + k, k \neq 0$.*

For instance, the whole-tone scale is equal to every one of its translates by an even number of semitones. Less trivially, a minor third is a combination of other minor thirds, as for instance

---

[4] There are 1,502 special cases out of 4,094 subsets of $\mathbb{Z}_{12}$, a fairly common occurrence.

$$(C, Eb) - (Eb, F\sharp) + (F\sharp, A) = (A, C)$$

This might appear to be a consequence of the minor third dividing the octave equally, but this is wrong, since the scale matrix of the major third is invertible, and the scale matrix of the fifth is singular.[5]

Since these singular cases are troublesome for reconstruction problems, [13] explored the simplest cases of singular subsets: dyads.

Theorem 3.8. *The pair* $(0, d)$ *in* $\mathbb{Z}_n$ *is never singular if* n *is odd. If* $n = 2^v q$ *with* q *odd, it is invertible iff* 2^v *divides* d, *the span of the dyad. Otherwise, the rank of the matrix associated with* $(0, d)$ *is equal to* $n - \gcd(d, n)$; *it is minimal for* $d = n/2$, *the equal division of the octave (the generalised tritone). In that case, it is equal to* $n/2$.

For instance, when $n = 12$ the only 'invertible dyad' is the major third.

Proof. The matrix $\mathscr{S}$ of the dyad $(0, d)$ is equal to identity plus the matrix $\mathscr{D}$ of the permutation $i \mapsto i + d \mod n$. Hence the kernel (or nullspace) of $\mathscr{S}$ is the eigenspace of $\mathscr{D}$ for eigenvalue -1. Let us reason geometrically, considering the vectors $e_0 \ldots e_{n-1}$ of the canonical basis of $\mathbb{R}^n$. A vector $x = \sum_{i=0}^{n-1} x_i e_i$ lies in this eigenspace iff

$$\sum_{i=0}^{n-1} x_i e_i = \sum_{i=0}^{n-1} -x_i e_{i+d} \iff \forall i = 0, \ldots n - 1 \quad x_{i+d} = -x_i$$

(all indexes are computed modulo n).

From this we get $x_{i+kd \mod n} = (-1)^k x_i$. Hence $x_{i+nd \mod n} = x_i = (-1)^n x_i$: if n is odd then the only solution is $x = 0$, i.e. $\mathscr{S}$ is invertible.

Say now that n is even, $n = 2^v q$ where q is odd. Let k be the smallest integer such that $kd = 0 \mod n$, e.g. $k = n/\gcd(d, n) = n/g$ (we put throughout $g = \gcd(d, n)$ for concision). If k is odd, for instance 2^v divides d, then we have the same impossibility, and S is invertible. We have proved that if 2^v divides d, then $\mathscr{S}$ is invertible.

Assume now that 2^v does not divide d, i.e. $d = 2^u d'$ with d' odd and $u < v$. We can produce the eigenvectors, i.e. elements of the kernel of S, in the following way:

- Fix one coordinate – say $x_0 = 1$.
- From the equation above, $x_d = x_{0+d} = -x_0 = -1$.
- Iterate until back to x_0: $x_{2d} = +1, x_{3d} = -1, \ldots x_0 = x_n = x_{n/g \times d} = +1$. The last value is indeed $+1$ because n/g is an even number.

So the value of one coordinate determines the value of n/g coordinates. We have thus $n/(n/g) = g$ arbitrary coordinates $x_0, x_1 \ldots x_{g-1}$, that is to say g degrees of freedom, and hence the dimension of the kernel of S is exactly g. Its largest possible value (apart from $d = 0$ which is no more a dyad) is for $d = n/2$. In general, we get the rank of matrix $\mathscr{S}$ by way of the rank-nullity theorem: $\operatorname{rank}(\mathscr{S}) = n - g$, remembering though that $\operatorname{rank}(\mathscr{S}) = n$ when 2^v divides d.

---

[5] The sum of all fifths beginning on one whole-tone scale is equal to the whole aggregate, as is the similar sum starting on the other whole-tone scale. Hence any single fifth is a linear combination of all the others.

The special case of the tritone (= half-octave) is worth a deeper analysis. Its matrix has the lowest possible rank, and more precisely all Fourier coefficients with odd index are nil. We can see for $n = 12$ how the codomain is generated by the first six columns, and the computation next to it shows the nullity of the odd Fourier coefficients.

$$\mathscr{T} = \begin{pmatrix} 1\,0\,0\,0\,0\,0\,1\,0\,0\,0\,0\,0 \\ 0\,1\,0\,0\,0\,0\,0\,1\,0\,0\,0\,0 \\ 0\,0\,1\,0\,0\,0\,0\,0\,1\,0\,0\,0 \\ 0\,0\,0\,1\,0\,0\,0\,0\,0\,1\,0\,0 \\ 0\,0\,0\,0\,1\,0\,0\,0\,0\,0\,1\,0 \\ 0\,0\,0\,0\,0\,1\,0\,0\,0\,0\,0\,1 \\ 1\,0\,0\,0\,0\,0\,1\,0\,0\,0\,0\,0 \\ 0\,1\,0\,0\,0\,0\,0\,1\,0\,0\,0\,0 \\ 0\,0\,1\,0\,0\,0\,0\,0\,1\,0\,0\,0 \\ 0\,0\,0\,1\,0\,0\,0\,0\,0\,1\,0\,0 \\ 0\,0\,0\,0\,1\,0\,0\,0\,0\,0\,1\,0 \\ 0\,0\,0\,0\,0\,1\,0\,0\,0\,0\,0\,1 \end{pmatrix} \qquad \widehat{t}(2p+1) = 1 + e^{\frac{(2p+1)6 \times 2i\pi}{12}} = 0.$$

This was actually noted by Yust in [96], who proved the following statement.

Lemma 3.9 (The tritone lemma). *Adding a tritone to a pc-set does not change its third and fifth Fourier coefficients.*

This follows directly from the linearity of the Fourier transform, and is also true for all other odd indexed coefficients. For instance, a fifth and the associated dominant seventh (GD and GBDF) have identical odd coefficients. So do a pentatonic (non hemitonic) scale and the associated diatonic (CDEGA and CDEFGAB), or even a single note and the diminished triad it divides (D and BDF). Conversely, one can remove a tritone from a melodic minor and get a singular hemitonic pentatonic with the same Fourier coefficients (ABCDEF♯G♯ → ABCEF♯). More impressive still, a minor triad has the same odd coefficients as the whole (harmonic minor) scale since they differ by two tritones. The most striking case I have found is the initial figure of Alban Berg's Sonata op. 1, which despite its spectacularly atonal character reduces to the single pc B when the tritones are removed, cf. Fig. 3.4.

There is a partial reciprocal, more technical, which involves Lemma 3.1.

Proposition 3.10. *Let A be a pc-set for which the Fourier coefficients $\mathscr{F}_A(3)$ and $\mathscr{F}_A(5)$ are nil. Then A is a tritone or a reunion of tritones.*

Proof. Consider the characteristic polynomial $\mathbf{A}(X) = \sum_{a \in A} X^a$.

Since the kth Fourier coefficient of A is simply $\mathscr{F}_A(k) = \mathbf{A}(e^{-2ik\pi/12})$ by Proposition 1.32, we are assuming that

$$e^{-2i3\pi/12} = -i \text{ and } e^{-2i5\pi/12} = e^{-5i\pi/6} = -\frac{\sqrt{3}}{2} + \frac{i}{2}$$

are roots of the polynomial $\mathbf{A}(X)$.

Mäßig bewegt.

Fig. 3.4. The initial motif and B have identical odd Fourier coefficients

$\mathbf{A}(X)$ has integer coefficients, the minimal polynomials of these roots in $\mathbb{Z}[X]$ are the cyclotomic polynomials $\Phi_4(X) = X^2 + 1$ and $\Phi_{12}(X) = X^4 - X^2 + 1$. Both being irreducible, $\mathbf{A}(X)$ must be a multiple of their product $\Phi_4(X) \times \Phi_{12}(X) = X^6 + 1$, which is the characteristic polynomial of a tritone.

Let $\mathbf{B}(X) = \dfrac{\mathbf{A}(X)}{X^6 + 1} = \sum_{0 \leqslant k \leqslant 5} b_k X^k$ be the exact quotient, with degree at most 5 since $\mathbf{A}(X)$ has degree at most 11.

It must have integer coefficients since $X^6 + 1$ is unitary, which must be 0's or 1's because they are coefficients of $\mathbf{A}(X)$:

$$\mathbf{A}(X) = (1 + X^6) \times \sum_{0 \leqslant k \leqslant 5} b_k X^k = b_0 + \dots b_5 X^5 + b_0 X^6 + \dots b_5 X^{11}.$$

Hence $\mathbf{A}(X)$ is the characteristic polynomial of a union of tritones, for example

$$(X^6 + 1) \times (X + X^2 + X^4) = X + X^7 + X^2 + X^8 + X^4 + X^{10}.$$

We leave as an exercise the generalisation to $\mathbb{Z}_n$ with even n.

One must beware that this does not exhaust all possible cases of non injectivity. For instance, as we will see in discussing the torus of phases, since dyads $\{0, 11\}$ and $\{4, 7\}$ have the same *phase* coordinates, so does their reunion, the major seventh $\{0, 4, 7, 11\}$.[6]   .

3.1.3 Structure of the zero set of the DFT of a pc-set

Lynx-eyed readers may have noticed that Lewin's conditions only consider the nullity of five Fourier coefficients. Perhaps this is sufficient because of the symmetry property $\mathscr{F}_A(n - k) = \overline{\mathscr{F}_A(k)}$, true for any real-valued distribution. Or is it? We left in the dark the values of $\mathscr{F}_A(5), \mathscr{F}_A(7)$. But actually it is enough to compute the $\mathscr{F}_A(k)$ when k is a divisor of n (in the set $\mathbb{N}$ of integers) because of the deep result below:

---

[6] I am indebted to J. Yust for this example.

Theorem 3.11. *For any rational-valued distribution f* (a fortiori *for any pc-set) we have*

$$\forall \alpha \in \mathbb{Z}_n^* \quad \widehat{f}(k) = 0 \iff \widehat{f}(\alpha k) = 0.$$

Remember that $\mathbb{Z}_n^*$ denotes the invertible elements of $\mathbb{Z}_n$. Other equivalent formulations involve *associated elements*:[7]

Definition 3.12. k *is associated with ℓ in $\mathbb{Z}_n$ $\iff$ $\exists \alpha \in \mathbb{Z}_n^*, \ell = \alpha k$.*

Actually the transformations $k \mapsto \alpha k$ for invertible α's are the automorphisms of the additive group $(\mathbb{Z}_n, +)$. Hence

Proposition 3.13.
- *Two elements of $\mathbb{Z}_n$ are associated iff they have the same order in the additive group $(\mathbb{Z}_n, +)$.*
- *Any element of $\mathbb{Z}_n$ is associated with (the class modulo n of) exactly one divisor of n.*
- *The classes of the relation 'being associated with' are the orbits of homotheties in $\mathbb{Z}_n$.*

For instance these classes in $\mathbb{Z}_{12}$ are $(0), (\mathbf{1}, 5, 7, 11), (\mathbf{2}, 10), (\mathbf{3}, 9), (\mathbf{4}, 8), (\mathbf{6})$. Thus Theorem 3.11 states that when the DFT vanishes in k it vanishes *for all classes modulo n associated with k*. Finally, this vindicates the exhaustiveness of the five Lewin's conditions, indexed by divisors 6, 4, 3, 2 and 1. The proof of the theorem involves cyclotomic polynomials again.

Proof. Let f be any integer-valued distribution[8] and $\mathbf{F} \in \mathbb{Z}[X]$ the associated polynomial: $\mathbf{F}(X) = \sum f(p)X^p$.

Say $\widehat{f}(k) = 0$. Since $\widehat{f}(k) = \mathbf{F}(e^{-2ik\pi/n})$ by Proposition 1.32, it means that $e^{-2ik\pi/n}$ is a root of $\mathbf{F}$. The order of $e^{-2ik\pi/n}$ in the group $(\mathbb{C}^*, \times)$ is $m = n/\gcd(n, k)$. By lemma 3.1, Φ_m must divide $\mathbf{F}$, hence all roots of unity with order m are roots of $\mathbf{F}$, i.e. all elements in $\mathbb{Z}_n$ associated with n/m are zeroes of the DFT, which is the result of the theorem.

It is high time we defined and considered the zero-set of a DFT:

Definition 3.14. *For a distribution $f \in \mathbb{C}^n$ (resp. a subset $A \in \mathbb{Z}_n$) the zero-set of its DFT is the set $Z(f)$ (resp. $Z(A)$) of the indexes k, satisfying $\widehat{f}(k) = 0$ (resp. $\mathcal{F}_A(k) = 0$).*

Theorem 3.11 proves that (for rational-valued distributions) $Z(A)$ is **structured as a reunion of classes** $d\mathbb{Z}_n^*$, orbits of associated elements, indexed by the set of divisors of n. Another way to put it is the invariance of $Z(A)$ under multiplication (by invertible elements). This is a strong feature: there are for instance $2^{20} - 1 = 1,048,575$

---

[7] Already met in the proof of Theorem 2.10.
[8] Actually this result is true for rational-valued coefficients, which is trivial in a way – any rational polynomial being an integer-coefficient polynomial divided by some integer – and deep too, because of the topological density of rational polynomials in $\mathbb{R}[X]$.

subsets of $\mathbb{Z}_{20}$, but only $64 = 2^6$ of them can be zero-sets, pieced together from six orbits which partition the whole group. This will provide access to a method of classification and exhaustive search for tiling canons as we will see in Section 3.3.

As we will develop soon, coverings with zero-sets is the condition for tiling by translation, and the relationships between the diverse classes constituting $Z(A)$ may give clues to abstract conditions for tiling and help lead to solutions of baffling open problems, such as the spectral conjecture.

Example 3.15.
1. For a tritone $T \subset \mathbb{Z}_{12}, Z(T) = \{1,3,5,7,9,11\}$.
2. For a melodic minor scale **mms** such as (A B C D E F♯ G♯) alias $\{0,2,4,6,8,9\}$, $Z(\mathbf{mms}) = \{2,10\}$.
3. Remember that in the example of $3-$homometry in $\mathbb{Z}_{32}$, one subset was

$$A = \{0,7,8,9,12,15,17,18,19,20,21,22,26,27,29,30\}.$$

Here $Z(A)$ is the set of even classes, which can be decomposed as

$$Z(A) = 2\mathbb{Z}_{32} \setminus \{0\} = \{2,4,\ldots,30\} = 2\,\mathbb{Z}_{32}^* \cup 4\,\mathbb{Z}_{32}^* \cup 8\,\mathbb{Z}_{32}^* \cup 16\,\mathbb{Z}_{32}^*.$$

4. Anticipating the next section, the subset $A = \{0,6,8,14\}$ tiles $\mathbb{Z}_{16}$, and

$$Z(A) = \{1,3,\mathbf{4},5,7,9,11,\mathbf{12},13,15\}$$
$$= \{1,3,5,7,9,11,13,15\} \cup \{\mathbf{4},\mathbf{12}\} = 1\,\mathbb{Z}_{16}^* \cup 4\,\mathbb{Z}_{16}^*.$$

Here the odd numbers are the invertibles (whose order is 16), and $\{4,12\}$ are the elements with order 4 in $\mathbb{Z}_{16}$ ($4 \times 4 = 12 \times 4 = 0$).

This is an algebraic constraint. One can well wonder how a Fourier coefficient manages to be equal to 0 in the first place. In the examples that we have detailed so far, it derived from Lemma 1.6, that it to say the exponentials involved in the sum are placed on the vertices of a regular polygon (for instance $1 + i + (-1) + (-i) = 0$ expresses the sum of the complex numbers on the vertices of a square). It seems natural to conjecture that, at least in the case of a subset distribution, a nil sum of exponentials can be decomposed into such regular subsums, a geometric constraint.

Conjecture 3.16. Let $A \in \mathbb{Z}_n$ such that $\sum\limits_{a \in A} e^{2i\pi a/n} = 0$. Then A can be partitioned as a disjoint reunion of regular polygons.

However, this conjecture is false as seen in Fig. 3.5.[9] The smallest counter-example that I found is $A = \{0,1,7,11,17,18,24\} \subset \mathbb{Z}_{30}$. Checking that the sum is exactly 0 involves finding the factor Φ_{30} in the characteristic polynomial $\mathbf{A}(X)$, see [14], which is equivalent to saying that $\mathbf{A}(e^{2i\pi/30}) = 0$. This sobering result warns that the study of nil Fourier coefficients is trickier than it seems.[10]

---

[9] Apparently it was first noticed in the 1950s but I could not find a precise reference. More about the algorithmic search for this counter-example in [14].

[10] A very recent paper [67] studies precisely those 'perfectly balanced sets' and hints that they can always be expressed as *algebraic linear combinations* of perfect polygons, in the

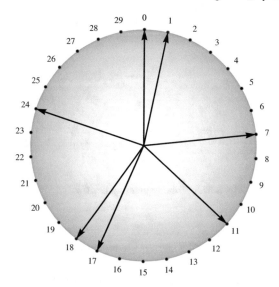

Fig. 3.5. Exponentials summing to 0 without regular subsums

3.2 Tilings of $\mathbb{Z}_n$ by translation

3.2.1 Rhythmic canons in general

The notion of musical canon is as old as the hills and remains popular even to our day in kindergarten songs. Informally, a canon is made of several voices playing the same tune, or pattern, or motif, at different times, i.e. starting with different offbeats. Often the canon is repeated in a loop and called a 'round', which expresses well its social function. Well-known examples in Anglo-Saxon culture are 'Brother John, Are You Sleeping?', 'Row, Row, Row Your Boat' or 'Three Blind Mice'. On the other hand, Ockeghem and Bach are known for brilliant intellectual constructions which played some part much later in the development of serial techniques.

Here we focus on just one musical dimension, usually considered as rhythm (though it could be any quantified musical quantity, and indeed there exist multi-dimensional canons tiling the spaces of rhythm and pitch for instance). Furthermore, in accordance with the topic of the book, we will mostly focus on canons by translations. It is of course possible to build canons with retrogradation, augmentation or any transformation of the motif, or to allow several notes to occur on the same beat (say an odd number of notes, see [27] for a recent study of canons mod p), but very little is known about these cases mathematically speaking (see [11] for a recent

---

spirit of linear combination of scales in [13]. For instance, my example can be decomposed as three pentagons: $\{0, 6, 12, 18, 24\} \oplus \{0, 1, 4\}$ united with two dyads $\{$or digons, or diameters$\}$ $\{2, 17\}, \{8, 23\}$ *minus* three dyads $\{1, 16\}, \{4, 19\}, \{10, 25\}$ and two equilateral triangles $\{0, 10, 20\} \oplus \{2, 8, 12\}$. This decomposition does prove the nullity of the Fourier coefficient. However it is hardly a practical method.

survey). A typical canon by translation is shown in Fig. 3.6 and was composed by George Bloch as a birthday greeting card (each voice sings 'Happy birthday').

Fig. 3.6. A birthday greeting periodic canon

The mathematical model of this canon is very simple: counting beats in sixteenth notes and setting the origin 0 at the start of the repeated bar, the four rhythmic voices are

$$\{0,4,5,9\},\{1,8,12,13\},\{2,3,7,14\},\{6,10,11,15\}$$

which are all copies of the initial $\{0,4,5,9\}$ with offsets of $0, 8, -2,$ and 6 respectively, the computation being made modulo 16 which expresses the repetition of a bar. A notion emerges: the tiling of a cyclic group with translates of one subset. Already we can see that the musical feature of repeating the bar models modular arithmetic. As we will see below, musical concepts are a great help in the mathematical study of rhythmic tilings.

Another essential feature of this canon is its perfect packing of the bar: each beat is played once and only once, which is a substantial difference from common musical canons where overlappings and silences are the rule rather than the exception. For musical treatment we will need this constraint (which still allows for billions of canons).

If only translations of the motif are allowed, it has been shown in the 1950s that a tiling of $\mathbb{Z}$ with a finite tile always has a period:

Theorem 3.17 (Hajòs, de Bruijn 1950). *Let A be a finite subset of $\mathbb{Z}$ and B such that $A \oplus B = \mathbb{Z}$. Then $\exists n \in \mathbb{N}^*, C \subset \mathbb{Z}$ such that $B = n\mathbb{Z} \oplus C$, i.e. $A \oplus C = \mathbb{Z}_n$ (reducing A, C modulo n).*

Hence the limitation to tilings of a cyclic group, which will be the only ones studied in this chapter. It has been recently shown by Kolountzakis and others [55] that the width of the motif does not really limit the period of the canon, refuting the long-standing conjecture that the latter was limited to twice the former (see again [11]).[11]

The study of tilings of cyclic groups (and more generally of abelian groups) was initiated in the 1950s, mostly by East-European mathematicians. The musical approach was single-handedly tackled by Dan Tudor Vuza ([94]) who rediscovered on

---

[11] The initial idea of Kolountzakis involves unfolding a cyclic group in 3 dimensions using its decomposition as a group product and geometric constructions. A similar vision probably presided over the creation of Szabó's counterexamples in [81], see Section 3.3.

his own the results of Hajòs, Redei, de Bruijn, Sands and others. The notion of 'Vuza canons' provided new impetus for these researches, especially since [6] connected them to difficult conjectures on tilings. Consequently, new algorithms have been devised for their enumeration ([57]), and these will be detailed below (section 3.3) for the sake of their relationship with DFT.

3.2.2 Characterisation of tiling sets

Definition 3.18. *A rhythmic canon*[12] *is a tiling of a cyclic group by translates of one tile, called motif.*

The motif A is called 'inner voice', and the set of its offsets is the 'outer voice' B. They form a rhythmic canon iff $A \oplus B = \mathbb{Z}_n$.

Example 3.19. In Fig. 3.6 one has $A = \{0,4,5,9\}, B = \{0,6,8,14\}$ and $A \oplus B = \{0,1,\ldots 15\} = \mathbb{Z}_{16}$.

Proposition 3.20.

$$A \oplus B = \mathbb{Z}_n \iff 1_A * 1_B = 1_{\mathbb{Z}_n} = 1$$

(the constant map equal to 1 for any element of $\mathbb{Z}_n$)

As we have seen in Chapter 1, the convolution product of characteristic functions turns into ordinary product of characteristic polynomials:

Proposition 3.21.

$$A \oplus B = \mathbb{Z}_n \iff \mathbf{A}(X) \times \mathbf{B}(X) = 1 + X + X^2 + \ldots X^{n-1} \mod (X^n - 1)$$

Either taking the DFT or plugging in $X = e^{-2i\pi k/n}$ in the last equation, we get

Proposition 3.22.

$$A \oplus B = \mathbb{Z}_n \iff \widehat{1_A} \times \widehat{1_B} = n\widehat{1_{\mathbb{Z}_n}} = n\delta = \left(x \mapsto \begin{cases} n & for\ x = 0 \\ 0 & else \end{cases} \right)$$

Essentially, setting apart the case of 0, the product of the Fourier transforms of the characteristic maps of the inner and outer voices must be nil. This vindicates again the definition of $Z(A) = \{k \in \mathbb{Z}_n, \widehat{1_A}(k) = 0\}$, the set of zeroes of the Fourier transform of (the characteristic map of) A already given above, and firmly grounds the question of tiling (by translation) in Fourier space:

Proposition 3.23. *Motif A tiles with outer voice B if and only if*

$$Z(A) \cup Z(B) = \mathbb{Z}_n \setminus \{0\} \quad and \quad \#A \times \#B = n.$$

The zeroes of the Fourier transforms of A and B must *cover* $\mathbb{Z}_n$ (minus 0), allowing overlaps. For instance, with $A \oplus B = \{0,4,5,9\} \oplus \{0,6,8,14\} = \mathbb{Z}_{16}$ (the factors in Fig. 3.6) we have

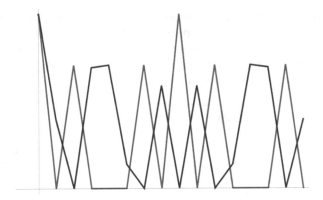

Fig. 3.7. $Z(A)$ and $Z(B)$ cover $\mathbb{Z}_{16} \setminus \{0\}$

$$Z(A) = \{1,3,4,5,7,9,11,12,13,15\} \quad \text{and} \quad Z(B) = \{2,6,8,10,14\}$$

as can be seen on the graphs of $|\mathscr{F}_A|$ and $|\mathscr{F}_B|$ featured in Fig. 3.7. Again, a complex phenomenon in musical space is seen at a glance in Fourier space, cf. Theorem 1.11.

At this point, the question of building all rhythmic canons with period n (i.e. all tilings of $\mathbb{Z}_n$ by translation, i.e. all factorisations $\mathbb{Z}_n = A \oplus B$), or the subproblem of 'completing' a given motif A with its counterpart B, appears as an extension of the phase retrieval problem: given a pair of zero sets covering $\mathbb{Z}_n$ – a very limited choice since these sets must be unions of a few orbits, according to Theorem 3.11 – is it possible to find corresponding subsets? But knowing only where $\mathscr{F}_A = 0$ is even less informative than knowing $|\mathscr{F}_A|$ (which is what we know in homometry questions) since the magnitude of the DFT has yet to be chosen where it is not (necessarily) nil; the problem is hence even more formidable. Precisely,

Proposition 3.24. *If A tiles $\mathbb{Z}_n$ with B (i.e. $A \oplus B = \mathbb{Z}_n$) then any A' homometric with A also tiles with B : $A' \oplus B = \mathbb{Z}_n$.*

This includes all the transforms of A under the dihedral group T/I, of course.[13] Less trivial cases are possible: for instance[14]

$$\text{both } A = \{0,1,6,10,12,13,15,19\}, \; A' = \{0,2,5,6,11,12,15,17\}$$
$$\text{tile } \mathbb{Z}_{24} \text{ with } B = \{0,8,16\},$$

though A, A' are homometric but not at all isometric (they both cover all residues modulo 8, however) as can be seen in Fig. 3.8.

---

[12] Properly speaking, a 'mosaic rhythmic canon by translation'.

[13] And there are scarcely any other sets homometric with a given A as seen in Chapter 2. This will be extended to affine transforms of A in Section 3.2.5.

[14] I am indebted to M. Andreatta who urged me to research these cases, probably the simplest subsets which tile and admit a non-trivially homometric twin.

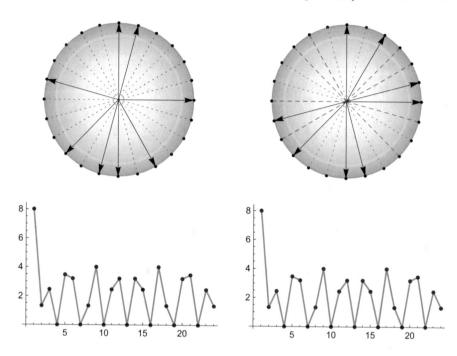

Fig. 3.8. Non-trivial homometric tiles, illustrating Proposition 3.24

However, some choices of $Z(A), Z(B)$ are impossible:[15] for instance a set like $A = \{0,2,4,5,6,7\}$ cannot possibly tile, as is easily gathered from trial and error (no way to fill the gaps 1, 3 with the 'lumpy' 4567 obstructing the process), and it can also be seen on $Z(A)$ as we will see in the next subsection. Notice that A tiles with its inversion $(3-A)$ though. Some reasons for such obstructions are known, and are our next topic.

3.2.3 The Coven-Meyerowitz conditions

[35] was the first paper enumerating general sufficient and (sometimes) necessary conditions for a finite motif to tile some cyclic group. Considering that the study of factorisations originated around 1948, this was long overdue. How does one check, for instance, whether $\{0,1,2,5,22,2415\}$ does tile[16], other than by finding a complement (which would be a long and arduous search considering the diameter of A)? Coven and Meyerowitz discovered that the cyclotomic factors of the characteristic polynomial are the key, and indeed provide something very close to a sufficient and necessary condition. As we have already explained, this prevalence of cyclotomic polynomials is another way of expressing the rigid structure of Fourier zero sets.

---

[15] For genuine pc-sets at least.

[16] It does. See below.

In [35] they introduced, for $A \subset \mathbb{Z}_n$,

Definition 3.25.
$R_A = \{d, d \mid n \text{ and } \Phi_d \mid A(X)\}$ and $S_A = \{p^\alpha \in R_A, p \text{ prime}, \alpha \geq 1\}$.

The elements of R_A are exactly the orders in $(\mathbb{Z}_n, +)$ of the elements of $Z(A)$, see Theorem 3.11:

$$Z(A) = \bigcup_{d \in R_A} \{x \in \mathbb{Z}_n \mid ord(x) = d\} = \bigcup_{d \in R_A} \frac{n}{d} \cdot \mathbb{Z}_n^*$$

For instance with $A = \{0, 3, 6, 12, 23, 27, 36, 42, 47, 48, 51, 71\}$ one gets $R_A = \{2, 8, 9, 18, 72\}, S_A = \{2, 8, 9\}$.[17]
The presence of all factors $\Phi_d, d \mid n$, in $A(X) \times B(X)$ entails that

- $S_A \cup S_B$ is the set[18] of all prime powers dividing n, and
- $R_A \cup R_B$ is the set of all divisors of n (1 excepted).

Coven and Meyerowitz then proceeded to prove the following statements, the last of which is quite difficult.

Theorem 3.26. *Defining conditions*

(T_1): $\prod_{p^\alpha \in S_A} p = \#A$;

(T_2): $p^\alpha, q^\beta, r^\gamma \cdots \in S_A \Rightarrow p^\alpha q^\beta r^\gamma \cdots \in R_A$ (*products of powers of distinct primes belonging to S_A are in R_A*);

one has

1. *If A tiles, then (T_1) is true.*
2. *If both $(T_1), (T_2)$ are true, then A tiles.*
3. *If #A has at most two different prime factors, and A tiles, then both $(T_1), (T_2)$ are true.*

As of today, it is not known whether condition (T_2) is always necessary for tiling. With the example above we can check (T_1) : $\#A = 12 = 2 \times 2 \times 3$ since $S_A = \{2^1, 2^3, 3^2\}$, and (T_2) : $2 \times 9 \in R_A$ and $8 \times 9 \in R_A$.

With the unreasonable tile given before, $A' = \{0, 1, 2, 5, 22, 2415\}$, with $\#A' = 6$ it is soon verified[19] that $S_{A'} = \{2, 3\}$ and $6 \in R_{A'}$, hence A' tiles quite trivially (it tiles $\mathbb{Z}_6$ and hence any $\mathbb{Z}_{6n}$).

---

[17] Actually the definition of [35] stands for $A \subset \mathbb{Z}$; we simplify slightly their exposition, since for any other polynomial congruent with $A(X) \mod (X^n - 1)$, the subset of the divisors of n in R_A, which are the indexes of the relevant cyclotomic factors, does not change. We choose this as our definition for R_A. Anyhow, S_A is always made of divisors of n.
[18] They show that corresponding cyclotomic polynomials occur only once, so this is a partition of the set of all prime powers dividing n. On the other hand, sometimes $R_A \cap R_B \neq \emptyset$.
[19] By computing $A(e^{2i\pi/3}) = 0 = A(-1)$.

Part of the proof of Theorem 3.26 is the useful

Lemma 3.27. *If A tiles some cyclic group, then it tiles $\mathbb{Z}_n$ where $n = \operatorname{lcm}(S_A)$ (reducing A modulo n).*

The link with Fourier transforms is straightforward: recall the organisation of $Z(A)$ in subsets of elements with equal multiplicative orders, these orders are precisely the elements of R_A.

3.2.4 Inner periodicities

Recall that $A \subset \mathbb{Z}_n$ is periodic, meaning $A + \tau = A$ for some $0 < \tau < n$, if and only if[20] $\mathscr{F}_A(t) = 0$ except when t belongs to some subgroup of $\mathbb{Z}_n$. This comes from $\mathscr{F}_{A+\tau}(t) = \mathscr{F}_A(t)e^{-2i\pi\tau t/n}$, hence $\mathscr{F}_{A+\tau}(t) = \mathscr{F}_A(t) \Rightarrow \mathscr{F}_A(t) = 0$ except when $\tau t \in n\mathbb{Z}$, i.e. $t \in \frac{n}{\gcd(\tau,n)}\mathbb{Z}$.

It turns out to be quite an effective way to check *a priori* periodicity, especially when one considers the *complement set of* $Z(A)$. The following theorem expresses the above in terms of $Z(A)$:

Theorem 3.28. *A is periodic in $\mathbb{Z}_n$ if and only if the complement set of $Z(A)$ is part of some subgroup of $\mathbb{Z}_n$. In practice, since any such subgroup is part of a maximal proper subgroup $p\mathbb{Z}_n$ with p a prime factor of n, it is sufficient to check whether there exists such a p which divides all elements* not *in $Z(A)$ in order to know whether A is n/p-periodic.*

This can be checked almost visually.

For $A' = \{0,5,8,13\}$, which tiles $\mathbb{Z}_{16}$, $R_{A'} = S_{A'} = \{2,(10),16\}$[21] and (keeping $n = 16$) the complement of $Z(A') = \{1,3,5,7,8,9,11,13,15\}$ is contained in the subgroup $2\mathbb{Z}_{16}$, meaning that A' is $16/2 = 8$-periodic. The non zeroes of the DFT are clearly members of the even subgroup materialised by big dots in Fig. 3.9 (though 8 is also a zero, inherited from $A = \{0,5\}$ from which A' is concatenated, see below).

In this example, $A' = A \oplus \{0,8\}$ where $A = \{0,5\}$, and we recognize the kinship between their respective Fourier transforms in Fig. 3.10. It is a 'multiplication d'accords' but in $\mathbb{Z}_{16}$, though the DFT of $\{0,5\}$ is drawn in $\mathbb{Z}_8$.

Some motifs can be completed by either periodic or aperiodic outer voices: $A = \{0,8,16,18,26,34\}$ tiles $\mathbb{Z}_{72}$ with $B = \{0,9,12,21,24,33,\ldots 60,69\} = \{0,9\} \oplus \{0,12,24,36,48,60\}$, 12-periodic, but also with $B' = \{0,3,12,23,27,36,42,47,48,51,71\}$. Comparison of zero sets is illuminating:

$$R_B = \{2,6,8,9,18,24,36,72\} \supset R_{B'} = \{2,8,9,18,72\}$$

---

[20] Notice that without loss of generality one may replace τ with $\gcd(\tau,n)$ and assume that τ is a divisor of n.

[21] Φ_{10} divides $\mathbf{A}(X)$ but is discounted since 10 is not a divisor of 16, according to Def. 3.25: this factor disappears if one changes any element of A' by a multiple of 16, see Footnote 17.

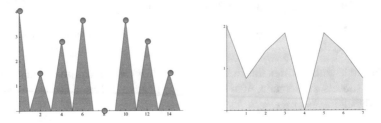

Fig. 3.9. The complement of $Z(A')$ is in $2\mathbb{Z}_{16}$, as seen on the graph of $|\mathscr{F}_{A'}|$. On the right, graph of $|\mathscr{F}_A|$.

Fig. 3.10. The Fourier transforms of $A = \{0,5\} \subset \mathbb{Z}_8$ and of $A' = \{0,5\} \oplus \{0,8\} \subset \mathbb{Z}_{16}$.

while $R_A = \{3,4,6,12,24,36\}$.

It is time to introduce

Definition 3.29. *A Vuza canon*[22] *is a counterexample to Hajós's 1950 conjecture, i.e. a rhythmic canon* $\mathbb{Z}_n = A \oplus B$ *where neither A nor B is periodic.*

I would like to point out that the notion of Vuza canons is musical, inasmuch as a canon with (say) a periodic outer voice is heard as the repetition of a shorter canon (with a shorter outer voice). This leads to a useful decomposition process, as we will see later. It took three decades for several top-notch mathematicians to establish the following theorem, which was rediscovered independently by D.T. Vuza in the 1980s ([77, 94]).

Theorem 3.30.
1. There exist Vuza canons.

---

[22] In some older papers, this term specifies those canons provided by Vuza's algorithm; this is no longer the case and we call 'Vuza canons' what he himself called 'Rhythmic Canons of Maximal Category'.

2. *Vuza canons exist in $\mathbb{Z}_n \iff n$ is* not *of the form*

$$n = p^\alpha, n = p^\alpha q, n = p^2 q^2, n = p^2 qr, n = pqrs$$

where p, q, r, s are different primes and $\alpha \geqslant 1$.

A cyclic group $\mathbb{Z}_n$ with n having any of the 5 forms above is often called, after Hajós, a 'good group'; the other cyclic groups are 'bad' (meaning that Hajós's conjecture fails in them). The smallest bad group is $\mathbb{Z}_{72}$, the next ones occur for $n = 108, 120, 144, 168, 180 \ldots$ [23] Classification of Vuza canons based on the zero sets of the factors is also a way of computing them exhaustively, which has been achieved for the values of n just stated. Some of the algorithms involved are mentioned in Section 3.2.3, the condition in Theorem 3.28 enabling the pruning of many cases where the only factors available would be periodic. The simplest construction of a Vuza canon in $\mathbb{Z}_n$ uses the recipe provided by Jedrzejewski: let p_1, p_2 be prime numbers and n_1, n_2, n_3 integers such that $\gcd(n_1 p_1, n_2 p_2) = 1$. Then have

$$A = n_2 n_3 \times \{0, \ldots p_2 - 1\} \oplus p_2 n_1 n_2 n_3 \times \{0, \ldots p_1 - 1\}$$
$$B = n_1 n_3 \times \{0, \ldots p_1 - 1\} \oplus p_1 n_1 n_2 n_3 \times \{0, \ldots p_2 - 1\}$$
$$S = p_2 n_2 n_3 \times \{0, \ldots n_1 - 1\} \oplus p_1 n_1 n_3 \times \{0, \ldots n_2 - 1\}$$
$$R = \left(\{1, \ldots n_3 - 1\} \oplus B\right) \bigcup A.$$

Then $R \oplus S = \mathbb{Z}_n$ yields a Vuza canon.

3.2.5 Transformations

Transformation of an existing canon has two obvious aims: the production of new canons, and their classification and taxonomy. For instance, $\{0, 4, 5, 9\}$ and its translate $\{0, 1, 5, 12\}$ tile identically $\mathbb{Z}_{16}$ with complement $\{0, 6, 8, 14\}$, itself the same as $\{0, 2, 8, 10\}$ if the origin of time is changed. Perceptively, in a canon repeated periodically, there is no privileged starting note or starting voice. Mathematically it is thus natural to consider the factors A, B up to translation in $\mathbb{Z}_n$. But there are other transformations which unravel less obvious relationships between canons.

Definition 3.31. *The dual canon of $A \oplus B = \mathbb{Z}_n$ is $B \oplus A = \mathbb{Z}_n$ (revert the roles of inner and outer voice).*

This is useful mainly for classification purposes, though some musical applications could be imagined. One other transformation does not change the size of the tiling:

Proposition 3.32. *If A tiles $\mathbb{Z}_n$ with B then mA tiles with B too for any m coprime with n.*

Proof. This is a direct consequence of Theorem 3.11, since the zero set $Z(mA)$ must be equal to $Z(A)$. Remarkably, this non-trivial feature of tilings was (re)discovered experimentally by not one, but several composers.

---

[23] Sloane's sequence of integers A102562.

This allows a finer classification of rhythmic canons than orbits under T or even T/I.

For instance, for $n = 72$ there are only two different Vuza canons up to affine transformation,

$$A = \{0,3,6,12,23,27,36,42,47,48,51,71\}$$
$$\text{or } A' = \{0,4,5,11,24,28,35,41,47,48,52,71\}$$

with one outer voice $B = \{0,8,10,18,26,64\}$ – instead of six inner voices and three outer voices under T/I.

Remember also that the famous Z-related sets $\{0,1,3,7\}$ and $\{0,1,4,6\}$ are affinely related[24] in $\mathbb{Z}_{12}$, but this is a more complicated case since non-nil Fourier coefficients must be permutated according to the affine transform. In this example, all odd (resp. even) coefficients share the same size $\sqrt{2}$ (resp. 2). A neater generalisation comes with J. Wild's FLIDs, see Section 4.3.3.

Further transformations of canons change n. In order to proceed we need to overcome an apparent ambiguity here: there is no canonical way to turn a subset of $\mathbb{Z}_n$ into a subset $\mathbb{Z}_{kn}$ but this will prove to be irrelevant:

Definition 3.33. *For any B in $\mathbb{Z}_n$, we call* immersion *of B in $\mathbb{Z}_{kn}$ any subset $B' \subset \mathbb{Z}_{kn}$ such that the canonical projection $\pi_n = \mathbb{Z}_{kn} \to \mathbb{Z}_n$ maps bijectively B' to B.*

In the transformations discussed below, any choice of B' will do, elements of B' being chosen up to a multiple of n.[25] The trick is to keep in mind that $R(B) = R(B')$ but $Z(B) \neq Z(B')$ when $\mathbb{Z}_n$ changes into $\mathbb{Z}_{kn}$. The rule is a simple one, preserving the multiplicative order:

Lemma 3.34. *With the same notations, $Z(B') = k(Z(B))$.*

The most important transformation is the next one:

Definition 3.35. *Concatenation of a canon consists in replacing the motif by itself, repeated several times. In other words, $A \in \mathbb{Z}_n$ turns into*

$$\overline{A}^k = A' \oplus \{0,n,2n,\ldots(k-1)n\} \in \mathbb{Z}_{kn}$$

where A' is an immersion of A.

For instance, $A = \{0,1,4,5\} \subset \mathbb{Z}_8$ (which tiles $\mathbb{Z}_8$ with $B = \{0,2\}$) can be prolonged to $\overline{A}^3 = \{0,1,4,5,8,9,12,13,16,17,20,21\} \subset \mathbb{Z}_{24}$. Obviously this new motif still tiles with complement $B' = \{0,2\} \subset \mathbb{Z}_{24}$. This is general:

Proposition 3.36. *A tiles $\mathbb{Z}_n$ with B if and only if $\overline{A}^k$ tiles $\mathbb{Z}_{kn}$ with B'.*

---

[24] In $\mathbb{Z}_{12}$, $5 \times \{0,1,3,7\} = \{0,3,5,11\} = \{0,1,4,6\} - 1$.

[25] In practice one uses the elements of B not caring whether they are integers, classes modulo n, or modulo kn, i.e. one replaces B with B' ruthlessly, usually choosing integers inside $[\![0,n-1]\!]$.

This property is easily checked with the geometric definition of a tiling[26], but with an eye on the next subsection, we will provide a more complicated proof involving the DFT.

Concatenation is the simplest recipe for building periodic motifs: $\overline{A}^k$ is n−periodic in $\mathbb{Z}_{kn}$, and conversely, any periodic motif is by nature concatenated from a shorter one. Hence as proved already, all Fourier coefficients, except those with index multiple of k, must be 0.

Lemma 3.37. *With the notations above, the elements of $Z(\overline{A}^k)$ have the same orders as those in $Z(A)$, plus those orders which are divisors of kn but not divisors of n:*

$$R(\overline{A}^k) = R(A) \cup \left(\mathrm{Div}(kn) \setminus \mathrm{Div}(n)\right).$$

This will entail Proposition 3.36, since all non-nil elements of $\mathbb{Z}_{kn}$ will fall either in $Z(\overline{A}^k)$ or $Z(B)$. Notice that elements with sthe ame orders are different because the group changes.

Proof. Using the characteristic polynomials:

$$\overline{\mathbf{A}^k}(X) = (1 + X^n + X^{2n} + \ldots X^{(k-1)n}) \times \mathbf{A}(X) = \frac{X^{kn} - 1}{X^n - 1} \times \mathbf{A}(X).$$

The roots of $\mathbf{A}(X)$ are still roots of $\overline{\mathbf{A}^k}(X)$, keeping the same order (as roots of unity), adding only the roots of $\dfrac{X^{kn} - 1}{X^n - 1}$, whose orders divide kn but not n, as stated.

Concatenation is an extension (to a larger group) of 'multiplication d'accords', i.e. a convolution product of characteristic functions or sum of (multi)sets: $\overline{A}^k = A' \oplus n\mathbb{Z}_{kn}$, and the computation of the zero set $Z(\overline{A}^k)$ might have been derived from the following trivial corollary of Theorem 1.10 (first noticed by J. Yust):

Proposition 3.38. *If a distribution f is singular (i.e. some Fourier coefficients are nil) then so is the convolution product $f * g$ for any distribution g. In terms of pc-(multi)sets, it means that if $A \subset \mathbb{Z}_n$ is one of Lewin's 'special cases,' then so is $A_{mult} + B_{mult}$ for any (multi)set B.*[27]

This is more general than the repetition/oversampling transformation that we have already considered in Chapter 1; it applies to collections of disjoint tritones or minor thirds, for instance. See Fig. 1.1 for an example of a singular set in Chopin which can be factored in a (singular) dyad × a (singular) minor triad.

Here is an example of computation of $Z(\overline{A}^k)$.

---

[26] $\overline{A}^k = A' \oplus n\mathbb{Z}_{kn}, A \oplus B = \mathbb{Z}_n, \mathbb{Z}_n' \oplus n\mathbb{Z}_{kn} = \mathbb{Z}_{kn} \Rightarrow \overline{A}^k \oplus B' = A' \oplus n\mathbb{Z}_{kn} \oplus B' = A' \oplus B' \oplus n\mathbb{Z}_{kn} = \mathbb{Z}_n' \oplus n\mathbb{Z}_{kn} = \mathbb{Z}_{kn}$.

[27] The index means that we consider multisets, and count multiplicities of elements of $A_{mult} + B_{mult}$ if necessary. Beware that this is different from the common (musicological) usage in 'multiplication d'accords' or transpositional combination.

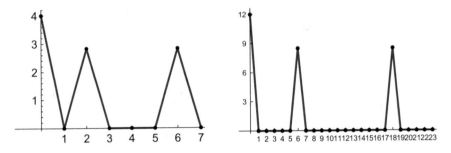

Fig. 3.11. Concatenation of $\{0,1,4,5\}$ and associated DFT.

Example 3.39. $A = \{0,1,4,5\}$ tiles $\mathbb{Z}_8$, the elements of $Z(A) \in \mathbb{Z}_8$ are $1,3,5,7$ and 4, with orders 8 and 2. For its third order repetition $\overline{A}^3 = A' \oplus \{0,8,16\} \subset \mathbb{Z}_{24}$, the elements of $Z(\overline{A}^3)$ have order $2,8$, and also the divisors of 24 which do not divide 8, i.e. $3,6,12$ and 24 – the multiples of 3 (see Fig. 3.11; the numerous new 0s are due to the additional orders, for instance $1, 5, 7\ldots$ have order 24). It is perhaps even more straightforward to look at the other side: $Z(B')$ is still made up of the elements with order 4, which were $2,6$ in $\mathbb{Z}_8$ and become $6,18$ in $\mathbb{Z}_{24}$ (the same, times 3).

This statement could also be expressed in terms of sets R_A and $R_{\overline{A}^{-k}}$ with R_A defined in Section 3.2.3 above, or alternatively with an expression of the DFT of a direct sum, a distinct possibility since a direct sum of subsets is 'une multiplication d'accords,' i.e. a convolution product of characteristic functions, i.e. a termwise product of DFTs. It could even be argued with sleight of hands that if a subset has some inner period, i.e. a smaller period than the size of the group it tiles, then fewer Fourier coefficient are required to describe the subset. The explicit description of the zero set that we have computed is a bit cumbersome but explicit.

Concatenation creates a periodic tile. Conversely, unless a canon is a Vuza canon, factor A or B (or both) is periodic, i.e. is a concatenation of smaller motifs. Iterating the process until it is no longer possible, we get the two following cases:

Proposition 3.40. *Any canon can be produced by concatenation (and duality) from either the trivial canon $\{0\} \oplus \{0\}$, or a Vuza canon.*

Moreover, this entails a recursive construction of *all* tilings of finite ranges $[\![0, n-1]\!]$ (i.e. without reduction modulo n), since

Theorem 3.41. *Any **compact** canon, i.e. $A \oplus B = [\![0, n-1]\!]$ (without reduction modulo n), can be reduced by concatenation and duality to the trivial canon.*

This was proved by N. G. de Bruijn in [37].

Example 3.42. $\{0,1,4,5\} \oplus \{0,2\} = [\![0,7]\!]$ is concatenated from $\{0,1\} \oplus \{0,2\} = [\![0,3]\!]$, this last from $\{0,1\} \oplus \{0\} = [\![0,1]\!]$ which is a duplication of the trivial canon $\{0\} \oplus \{0\} = [\![0,0]\!]$.

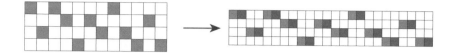

Fig. 3.12. Example of stuttering

Other cases of reducible canons include the 'assymmetric rhythms' of [48], whose study originates in ethnomusicology.

Zooming and stuttering are two dual transformations. I called *stuttering* (one could see it as 'upsampling') the act of replacing each note or rest in the motif by k repetitions of itself. Of course one must again replace $\mathbb{Z}_n$ with $\mathbb{Z}_{kn}$ in the process.

Example 3.43. From $\{0,2,7\} \oplus \{0,3,6,9\} = \mathbb{Z}_{12}$ one gets

$$\{(0,1),(4,5),(14,15)\} \oplus \{0,6,12,18\} = \mathbb{Z}_{24},$$

cf. Fig. 3.12.

Algebraically, this means turning A into[28] $Stut(A,k) = (kA)' \oplus \{0,1,2\dots k-1\} \subset \mathbb{Z}_{kn}$ (remember that $(kA)'$ is kA seen in $\mathbb{Z}_{kn}$). This time, in order to keep a canon it is necessary to *augment*, i.e. zoom in, the outer voice B into kB', i.e.

Theorem 3.44. *A tiles $\mathbb{Z}_n$ with B if and only if $Stut(A,k) = (kA)' \oplus \{0,1,2\dots k-1\}$ tiles $\mathbb{Z}_{kn}$ with kB'.*

In this book, we find it desirable to clarify what happens to the DFT during such transformations.

Lemma 3.45. *The transformation $B \mapsto kB'$ from $\mathbb{Z}_n$ to $\mathbb{Z}_{kn}$ turns $Z(B)$ into $Z(kB') = Z(B)' \oplus n\mathbb{Z}_{kn}$. Equivalently, $R(kB') = R(B)$.*

Proof. This is what we had already stated about oversampling. In the following line, t' is any preimage in $\mathbb{Z}_{kn}$ of $t \in \mathbb{Z}_n$, i.e. $t = \pi_n(t')$, i.e. $t' \equiv t \mod n$:

$$\mathscr{F}_{kB}(t') = \sum_{x \in kB} e^{-2i\pi xt'/(kn)} = \sum_{y \in B} e^{-2i\pi yt'/n} = \sum_{y \in B} e^{-2i\pi yt/n} = \mathscr{F}_B(t)$$

does not change with the choice of t', i.e. if t' is modified by some multiple of n. Hence $\mathscr{F}_{kB}$ vanishes on $Z(B) \oplus n\mathbb{Z}_{kn}$.

Example 3.46. Say $B = \{0,1,4,5\} \subset \mathbb{Z}_8$, then $Z(B) = \{1,3,4,5,7\}$ and

$$Z(3B)' = \{1,3,4,5,7,\,9,11,12,13,15,\,17,19,20,21,23\} = \{1,3,4,5,7\} \oplus \{0,8,16\}.$$

---

[28] It may be construed as a kind of tensorial product, as Franck Jedrzejewski showed in an unpublished conference at the MaMuX seminar in IRCAM (Paris). With the matricial formalism introduced in Section 1.2.3, this is equivalent to tensorial products of matrixes, which would yield the same results but in a more cumbersome way.

Lemma 3.47. *Stuttering A into* $Stut(A,k) = (kA)' \oplus \{0,1,2\ldots k-1\} \in \mathbb{Z}_{kn}$ *from* $\mathbb{Z}_n$ *to* $\mathbb{Z}_{kn}$ *turns* $Z(A)$ *into* $Z(Stut(A,k)) = n\mathbb{Z}_{kn} \cup (Z(A) \oplus n\mathbb{Z}_{kn}) \setminus \{0\}$.

Proof.

$$\mathscr{F}_{Stut(A,k)}(t') = \sum_{x \in kA' \oplus [[0,k-1]]} e^{-2i\pi x t'/(kn)} = \sum_{a \in A, \ell \in [[0,k-1]]} e^{-2i\pi(ka+\ell)t'/(kn)}$$

$$= \sum_{a \in A} e^{-2i\pi a t'/n} \times \sum_{\ell \in [[0,k-1]]} e^{-2i\pi \ell t'/(kn)} = \mathscr{F}_A(t) \times \begin{cases} \dfrac{1-e^{-2i\pi t'/n}}{1-e^{-2i\pi t/(kn)}} & \text{if defined} \\ k & \text{else.} \end{cases}$$

This is 0 whenever t' is a multiple of n or when $t \in Z(A)$, i.e. $t' \in Z(A) \oplus n\mathbb{Z}_{kn}$.

Example 3.48. Let $A = \{0,2,7\} \in \mathbb{Z}_{12}$: since $\mathbf{A}(e^{2i\pi/3}) = 1 + j^2 + j^7 = 0$, $Z(A) = \{4,8\}$. Now for $3A' \oplus \{0,1,2\} = \{0,1,2, 6,7,8, 21,22,23\}$ we get the zero set

$$\{4,8,12,16,20,24,28,32\} = (\{0,4,8\} \oplus \{0,12,24\}) \setminus \{0\}.$$

Quite contrary to concatenation, these operations preserve the non-periodicity of either voice, and hence turn a Vuza canon into a (larger) Vuza canon. Historically, this has been used (in combination with the other transformations) in order to produce larger Vuza canons, for instance before Harald Fripertinger managed to enumerate all of them for periods 72 and 108 ([44]). Of course, it is equally possible to zoom on A and stutter with B.

Multiplexing is a generalisation of stuttering (see example in Fig. 3.13): instead of building $\{0,1,2\ldots k-1\} \oplus kA$, one chooses k inner voices $A_0,\ldots A_{k-1}$ which tile with the same outer voice B, i.e. $A_0 \oplus B = A_1 \oplus B = \cdots = \mathbb{Z}_n$, and the new motif with period kn is $\widetilde{A} = \bigcup_{i=0}^{k-1} (kA_i' + i)$. Again,

Theorem 3.49. $\widetilde{A} \oplus kB' = \mathbb{Z}_{kn} \iff \forall i = 0\ldots k-1, A_i \oplus B = \mathbb{Z}_n$.

The easy proof is left to the reader.

It seems ambitious to look for the zero set of such a complicated construction. But all $Z(A_k)$'s have enough in common to warrant a statement:

Lemma 3.50. $Z(\widetilde{A})$ *is at least the same as the zero set obtained by stuttering,*

$$Z(\widetilde{A}) \supset Z(Stut(A,k)) = n\mathbb{Z}_{kn} \cup (Z(A) \oplus n\mathbb{Z}_{kn}) \setminus \{0\} = ((\{0\} \cup Z(A)) \oplus n\mathbb{Z}_{kn}) \setminus \{0\}$$

(according to Lemma 3.47) if we define $Z(A)$ *as* $\cap_k Z(A_k)$, *which complements* $Z(B)$ *in* $\mathbb{Z}_n$ *by hypothesis.*

Example 3.51. In Fig. 3.13, both motifs $\{0,1,11\}, \{0,2,7\}$ share the same $Z(A) = \{4,8\}$ and hence the multiplexed motif satisfies $Z(\widetilde{A}) \supset \{4,8,12,16,20\}$.

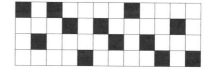

$$\{0,1,11\} \oplus \{0,3,6,9\} = \mathbb{Z}_{12}$$

$$\{0,2,7\} \oplus \{0,3,6,9\} = \mathbb{Z}_{12}$$

$$\big(2 \times \{0,1,11\} \cup 2 \times (\{0,2,7\}+1)\big) \oplus 2 \times \{0,3,6,9\}$$
$$= \{0,1,2,5,12,22\} \oplus \{0,6,12,18\} = \mathbb{Z}_{24}$$

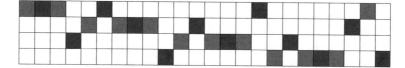

Fig. 3.13. An example of multiplexing.

This transformation opens interesting compositional possibilities, since several canons merge into a larger one while remaining audible, cf. Fig. 3.13. The dual transformation (multiplexing the outer voice) enlarges the motif and complexifies its outer voice.

An interesting theoretical point is that a kind of reciprocal stands: each canon wherein the outer voice can be written kB (i.e. up to translation, all elements of the outer voice are divisible by a common k) is multiplexed from a canon k times smaller (see in Fig. 3.13 how the smaller canons can be retrieved from the larger one). It was conjectured, in various contexts and by several authors, that essentially all canons were instances of some such multiplexing; but this is not true, as demonstrated by [81], though the smallest known counter examples have period 900, see Section 3.3. This precludes, to this day, reducing all canons to the trivial canon.

Uplifting

The last transformation we will study, uplifting (Fig. 3.14), came to the fore in recent developments of the search for Vuza canons [57], though it was probably used by composers before. It stems from the simple idea that allowed us above to immerse a subset of $\mathbb{Z}_n$ in a larger group:

Proposition 3.52. *If A tiles $\mathbb{Z}_n$ then A – or rather its immersion A' – tiles any larger cyclic overgroup $\mathbb{Z}_{kn}$; moreover, translating any individual element of A' by any multiple of n provides a new motif A'' that also tiles $\mathbb{Z}_{kn}$.*

Proof. If $A \oplus B = \mathbb{Z}_n$, let $A' = \{a_1 + k_1 n, \ldots a_p + k_p n\} \subset \mathbb{Z}_{kn}$ where $A = \{a_1, \ldots a_p\} \subset \mathbb{Z}_n$ and $k_1, \ldots k_p \in \mathbb{Z}$. This makes sense, since applying the canonical projection π_n from $\mathbb{Z}_{kn}$ to $\mathbb{Z}_n$ yields $\pi_n(a + kn) = a$ as in the other transformations studied above. Let also $B' = \{b_i + \kappa n, b_i \in B, \kappa = 0, \ldots k - 1\}$; then it is straightforward to check that

$$A' \oplus B' = \mathbb{Z}_{kn},$$

considering[29] that the map $A' \times B' \ni (a,b) \mapsto a+b$ is still injective and that $\#A' \times \#B' = kn$.

Again, one can reach most of the zero set of the new motif:

Lemma 3.53. $Z(A')$ *contains at least* $kZ(A)'$ *(equivalently, $R(A') \supset R(A)$).*

Proof. When A is immersed in $\mathbb{Z}_{kn}$ its DFT changes and $Z(A)$ (in $\mathbb{Z}_n$) turns into $Z(A') \supset kZ(A)'$ (in $\mathbb{Z}_{kn}$), since now

$$t' \in kZ(A)' \Rightarrow \mathscr{F}_A(t') = \sum_{a' \in A'} e^{\frac{-2i\pi a't'}{kn}} = \sum_{a' \in A'} e^{\frac{-2i\pi a'kt}{kn}} = \sum_{a \in A} e^{\frac{-2i\pi at}{n}} = \mathscr{F}_A(t) = 0$$

since $t' = kt$ for some $t \in Z(A)$.

As we see in the computation, changing any element $a \in A$ by any multiple of n does not change the result.

For instance, from $\{0,1,4,5\} \oplus \{0,2\} = \mathbb{Z}_8$ one 'uplifts' to the Bloch canon in Example 3.6, e.g.

$$\{0,9 = 1+8,4,5\} \oplus \{0,2,8,10\} = \mathbb{Z}_{16}.$$

The zero sets are respectively $\{1,3,4,5,7\} \subset \mathbb{Z}_8$ and $\{2,6,8,10,14\} \subset \mathbb{Z}_{16}$, the orders of their elements being in both cases 8 or 2.

This is most probably what Bloch actually did in order to produce his canon. But the main strength of this transformation is made clear when one is looking for some motif $A \in \mathbb{Z}_{kn}$ *knowing that A also tiles $\mathbb{Z}_n$.* This was instrumental in many cases in the algorithmic quest for all the smallest Vuza canons, see [57, 11] and Section 3.3 below.

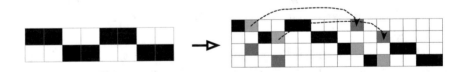

Fig. 3.14. Uplifting a canon in $\mathbb{Z}_8$ to a larger one in $\mathbb{Z}_{16}$

In all these transformations, we keep control of $Z(A)$. Hence, in order to prove most conjectures on rhythmic canons, it is enough to check only those canons who generate all other ones by those transformations, i.e. Vuza canons.

---

[29] Alternatively one can reason on sets, writing $B' = B \oplus n\mathbb{Z}_{kn}$.

3.2.6 Some conjectures and routes to solve them

The (T_2) conjecture

Let us recall the Coven-Meyerowitz conditions introduced in Section 3.2.3:

1. If A tiles, then (T_1) is true.
2. If both $(T_1), (T_2)$ are true, then A tiles.
3. If #A has at most two different prime factors, and A tiles, then both $(T_1), (T_2)$ are true.

[35] carefully refrained from enunciating the sometimes improperly stated 'Coven-Meyerowitz conjecture,' namely

Conjecture 3.54. A tiles $\iff$ both $(T_1), (T_2)$ are true.

The discussion of $Z(A)$ in the section above shows that (T_2) is inherited through all transformations:

Theorem 3.55. *If $A \oplus B = \mathbb{Z}_n$ is concatenated (or zoomed, or stuttered, or multiplexed) to a larger rhythmic canon, then (T_2) is true for the large canon whenever it is true for the smaller one.*

Proof. Consider for instance $\overline{A}^k$, the concatenation of the motif A k times. As we have established above, $R(\overline{A}^k) = R(A) \cup (\text{Div}(kn) \setminus \text{Div}(n))$. Hence $S(\overline{A}^k)$ is $S(A)$, adding p^α whenever $p^\alpha \mid k$ though p does not divide n, and changing p^α to $p^{\alpha+\beta}$ if p^α, p^β are the powers of p in n, k respectively.

Remark 3.56. Checking condition (T_1) was not required, because it must be satisfied in both the short and large canons; but it would be straightforward to verify that it is true for A whenever it is true for $\overline{A}^k$.

From the equation above, clearly condition (T_2) holds in $R(\overline{A}^k)$ iff it holds in $R(A)$: apart from $R(A)$ itself, in $R(\overline{A}^k)$ we have also all terms with p^α as a factor when p divides k but not n, and when p is a factor of both k and n then the $p^{\alpha+\beta} q^\gamma \ldots$ as above are in $R(\overline{A}^k)$ since they are divisors of kn **but not of n** because the exponent of p is too large.

The other factor B of the tiling does not change[30], and neither do $R(B), S(B)$ or hence condition (T_2) for B. The proof is similar for other transformations, using the results of the lemmas in last section.

Similar arguments hold for the other transformations, see [6, 46]. Since any canon can be deconcatenated down to a Vuza canon (or to the trivial canon, $\{0\} \oplus \{0\}$), it follows:

Proposition 3.57. *Conjecture 3.54 is true $\iff$ it is true for Vuza canons.*

---

[30] With the notations above, B changes to B' but for instance the polynomial $\mathbf{B}(X)$ stays the same.

This result revived the interest in Vuza canons when it was first published in [6], proving Conjecture 3.54 (and the spectral conjecture below too) in 'good groups', adding cases $n = p^m qr, pqrs$ (with p,q,r,s distinct primes) to [35]'s case $n = p^\alpha q^\beta$.

This deconcatenation technique also applies to all 'compact canons' (i.e. $A \subset \mathbb{Z}, B \subset \mathbb{Z}$ with $A \oplus B = \{0,1,2\ldots n-1\}$ without modulo n reduction), and [35] already noted that this implied the truth of Conjecture 3.54 in that case.

The spectral conjecture

Despite its name, the origin of the *spectral conjecture* is extraneous to the field of the present book, but it is still open in dimension 1 and 2, the former being our topic. It states

Conjecture 3.58. (Fuglede, 1974) *A tiles some* $\mathbb{Z}_n \iff A$ *is spectral.*[31]

Here, 'spectral' means that the tile (a measurable subset of $\mathbb{R}^n$ in the most general context) admits a Hilbert basis of exponential functions, meaning, in the seminal case, that any map in $L^2([0,1[)$ is the sum of its Fourier expansion. In dimension 1 we have a less esoteric definition involving difference sets:

Definition 3.59. *A subset* $A \in \mathbb{Z}$ *is spectral if there exists a* spectrum $\Lambda \subset [0,1[$, *i.e. a subset with the same cardinality as A, such that* $e^{2i\pi(\lambda_i-\lambda_j)}$ *is a root of the characteristic polynomial* $\mathbf{A}(X)$ *for all distinct* $\lambda_i, \lambda_j \in \Lambda$.

In other words, $Z(A)$ must include a (large enough) difference set.

It is still unknown whether in general the $\lambda_i - \lambda_j$ must be rational, i.e. whether the roots in question are roots of unity, though some progress was recently made in that respect. But if we consider A as a set in $\mathbb{Z}$ defined modulo $n\mathbb{Z}$, i.e. any element of A can be twiddled by any multiple of n – since this does not change the condition that A tiles $\mathbb{Z}_n$ – then *only those roots of* $A(X)$ *which are* n^{th} *roots of unity are unchanged*. Hence we may assume that $\Lambda \subset \{0, \dfrac{1}{n}, \dfrac{2}{n} \ldots \dfrac{n-1}{n}\}$, i.e. $n\Lambda \subset \mathbb{Z}_n$, which we will do henceforth.[32]

The spectral conjecture has been proved in many cases (convex tiles for instance) but in general it is false, as first shown in high dimension by Fields medalist Terence Tao [82]. Following further work [56], the conjecture only remains open in dimensions 1 and 2. In dimension 1, which is our context for rhythmic canons, Izabella Łaba has proved [59] that $(T_1) + (T_2)$ implies 'spectral', explicitly constructing a spectrum under these conditions, just as [35] proved that $(T_1) + (T_2)$ implies 'tiling'. So the conjecture is known to be true when n has only two prime factors, by the

---

[31] Originally it is a question of tiling $\mathbb{R}^n$ but in dimension 1 it can be reduced to tilings of $\mathbb{Z}$, see [45, 59].

[32] Twiddling an element by n adds $X^n - 1$ to the characteristic polynomial $\mathbf{A}(X)$, which destroys any root which is not common to both polynomials, hence this statement. [46] argues for this restricted definition of 'spectral', through characters of the group $\mathbb{Z}_n$, which also makes perfect sense and yields the same overset of Λ. Perhaps this condition should be properly labeled 'spectrality in a cyclic group'.

last result in Theorem 3.26; it is also true for motifs that tile a 'good group', because by deconcatenation such a tiling reduces to the trivial tiling and hence inherits $(T_1) + (T_2)$ (first proved in [6]). More generally, it is true for any motif that can be reduced to a tiling satisfying $(T_1) + (T_2)$, for instance the compact tilings mentioned above.[33]

Without condition (T_2) we have a direct heredity result:

Theorem 3.60. *Let $A \subset \mathbb{Z}$ be a finite motif of some tiling. We know from [35] that it tiles $\mathbb{Z}_n$ with $n = \gcd(R(A))$; then $\overline{A}^k$ is spectral if and only if A is spectral.*

This was announced in [3, 8], but first properly stated and proved in printed form in [46], which we follow below. If all Vuza canons are spectral, meaning both factors A, B are spectral sets, then by concatenation (and duality) any canon is spectral too. Hence the spectral conjecture (in the direction tiling $\Rightarrow$ spectral) is true if and only if it is true for all Vuza canons, which is another stringent motivation for their study.

Proof. Consider the concatenation of $A, \overline{A}^k \subset \mathbb{Z}_{kn}$. We have proved above that $R(\overline{A}^k) = R(A) \cup \big(\mathrm{Div}(kn) \setminus \mathrm{Div}(n)\big)$. Assume that we know a spectrum Λ for A, meaning that $e^{2i\pi(\lambda_i - \lambda_j)}$ is a root of the characteristic polynomial $\mathbf{A}(X)$ for all distinct $\lambda_i, \lambda_j \in \Lambda$. But in the ring of polynomials,

$$\overline{\mathbf{A}}^k(X) = (1 + X^n + X^{2n} + \ldots X^{(k-1)n}) \times \mathbf{A}(X) = \frac{X^{nk} - 1}{X^n - 1} \times \mathbf{A}(X).$$

Hence Λ already produces some roots of $\overline{\mathbf{A}}^k(X)$. But $\#\overline{A}^k = k \times \#A$ and we need a larger spectrum. A possible solution is the sum

$$\Lambda' = \Lambda + \left\{0, \frac{1}{nk}, \frac{2}{nk} \ldots, \frac{k-1}{nk}\right\}.$$

First, this spectrum has the right cardinality $k\#A$ (one has to check that the sum is direct, this follows from the fact that $\lambda_i - \lambda_j = q/n$ as assumed above).

Last, any element of Λ', i.e. $(\lambda_i - \lambda_j) \pm \dfrac{p}{nk}$, is equal either to $\lambda_i - \lambda_j$ (when $p = 0$), providing a root of $\mathbf{A}(X)$ as mentioned in the beginning of the proof, or to some $\dfrac{q}{n} \pm \dfrac{p}{nk}$ with $-n < q < n$ and $-k < p < k$, and hence provides a root of $X^{nk} - 1$ which is not a root of $X^n - 1$, i.e. one of the additional roots in $R_{\overline{A}^k}$. In both cases we get a root of $\overline{\mathbf{A}}^k(X)$ and hence Λ' is a spectrum.

For the complete reduction of Fuglede's conjecture to Vuza canons (or to the trivial canon when the deconcatenation process only ever stops with $\{0\} \oplus \{0\}$), one also needs the preservation of the spectral condition under duality (exchanging the

---

[33] In some cases I was able to predict that any Vuza canon in $\mathbb{Z}_{180}$ with a specific value of R_A could be reduced by demultiplexing to a canon with period 90, implying (T_2), without finding explicitly the canons in question but knowing from the factors in R_A that any complement B of A would be divisible by 2, i.e. that the canon could be demultiplexed.

factors) and prolongation of the other factor in concatenation $B \subset \mathbb{Z}_n$ to $\mathbb{Z}_{kn}$, which are both trivial, as is the zooming operation changing $A \subset \mathbb{Z}_n$ into $kA' \subset k\mathbb{Z}_n$. Gilbert also proved that the condition is invariant under affine transform or even multiplexing under conditions analogous to our computation above, when the new $\mathbb{Z}_{\tilde{A}}$ is computed from the $\bigcap \mathbb{Z}_{A_i}$ (assuming this intersection is spectral in a natural sense).

Example 3.61. Consider $A = \{0,1,4,5\}$, $R_A = \{2,8\}$ and hence A tiles $\mathbb{Z}_n$ for $n = 8$. Triple concatenation of A yields

$$\overline{A}^3 = \{0,1,4,5,8,9,12,13,16,17,20,21\}$$

and $R_{\overline{A}^3}$ is made of all integers below 24 except 6 and 18 (Fig. 3.11). This happens because $\overline{A}^3(X)$ is a pure product of cyclotomic polynomials:[34]

$$\overline{A}^3(X) = \Phi_2 \Phi_3 \Phi_6 \Phi_8 \Phi_{12} \Phi_{24}$$

A spectrum for A is [35] $\Lambda = \{0, \frac{1}{2}\} \oplus \{0, \frac{1}{8}\} = \{0, \frac{1}{8}, \frac{1}{2}, \frac{5}{8}\}$.[36] For a spectrum in $\overline{A}^3$ one adds $\frac{0,1,2}{24}$ and finally $\Lambda' = \frac{1}{24}\{0,1,2,3,4,5,12,13,14,15,16,17\}$ with 12 elements as required, whose differences yield all values of $\frac{k}{24}$ barring $\frac{6}{24}$ and $\frac{18}{24}$, as desired.

Detailed algorithms are provided in Section 3.3.

3.3 Algorithms

3.3.1 Computing a DFT

The definition formula is easy to implement in any modern programming language: loop over both the elements of the pc-set and the indexes. In the most general case, for a distribution $f \in \mathbb{C}^{\mathbb{Z}_n}$ one

- Selects (or input) the index k of the coefficient.
- Sets $s = 0$.
- For j from 0 to $n - 1$, does $s = s + e^{-2i\pi kj/n} \times f(j)$.
- Returns the value of s: it is $\widehat{f}(k)$.

---

[34] Building up rhythmic canons from products of cyclotomic polynomials was tried in [1] and implemented in *OpenMusic*. It is a fairly quick process – list cyclotomic polynomials, select index lists satisfying condition (T_2) and effectuate the corresponding product, discard the result if it is not 0-1, else find the possible outer voices – but omits many canons.

[35] I follow Łaba's recipe in [59].

[36] Search for a spectrum may well require exponential time, unless conjecture 3.54 is true, since the Coven-Meyerowitz conditions can be checked in polynomial time, as pointed out by Kolountzakis.

At worst one can separate real and imaginary parts and compute them separately (the former a sum of cosines, the latter a sum of sines).

Using $\cos(\pi/6) = \sqrt{3}/2, \sin(\pi/6) = 1/2$ and other trigonometric values, one can even compute a DFT by hand (preferably beginning with the kind of geometrical simplifications suggested in Fig. 3.3). Some practical advice: numerical calculations often fail to identify 0, so a routine that tidies the results (turning any $x \in [-10^{-10}, 10^{10}]$ to 0 for instance) is generally a good idea, especially for inverse Fourier transform.

Many high-level environments will provide a ready-made Fourier transform. One has to check which convention is used and perhaps adjust the result. For instance in Mathematica™, the DFT of a pc-set (say $\{0,4,7\}$) as defined in this book could be obtained with the native function **Fourier** by

Fourier $[\{1,0,0,0,1,0,0,1,0,0,0,0\},$ **FourierParameters** $\rightarrow \{1,-1\}]$.

Notice that the pc-set is replaced by the associated distribution – this can be automated by something like

Table[If [MemberQ[set, k], 1, 0], $\{k,0,n-1\}]$

unless one prefers to compute one's own DFT with a loop inside a loop, as described above.

Major Scale Similarity

I include in this subsection the computation of **Major Scale Similarity** (MSS) though it is only defined below. One has to input first a temperament (TeT). Say it is given as a table of values in cents – for instance, Werkmeister's fifth TeT is

$$(0, 107.8, 209.8, 305.9, 407.8, 503.9, 611.7, 707.8, 803.9, 911.7, 1007.8, 1109.8).$$

Now define the magnitude of the first Fourier coefficient of a scale[37] (i.e. a table of 7 values in cents) as

$$A(\text{scale}) = \sum_{k=0}^{6} e^{2i\pi \, \text{scale}[k]/1200} e^{-2ik\pi/7}$$

(beware of your programming language's conventions; here I assume that the first index of a table is 0).

Compute the table of all major scales in the given TeT: starting from the list of indexes $ind = [0,2,4,5,7,9,11]$, run the 12 transpositions, i.e. $ind + k \pmod{12}$, and tabulate

$$\text{scale}[k] = \text{table}(\text{TeT}[(\text{ind}[j] + k) \pmod{12}], j = 0\ldots6).$$

With a simple loop, compute the max M and min m of the 12 values $A(\text{scale}[k])$:

- $m = 1000, M = 0$.

---

[37] With Noll's order-dependent definition, see $\mathfrak{F}_{\mathscr{A}}(1)$ inSection 5.2.

- For $k = 0$ to 11, do $x = A(\text{scale}[k])$;
    - If $x > M$ then $M = x$;
    - If $x < m$ then $m = x$;

Now the value of MMS(TeT) is $\dfrac{1}{M - m}$.

3.3.2 Phase retrieval

For convenient reference, I repeat here the algorithm for finding the unknown coefficient in Lewin's problem when one Fourier coefficient is nil:

1. Compute the cardinality of A: it is the sum of the elements of $\text{IFunc}(A,B)$ divided by #B.
2. Compute $\mathscr{F}_A = \dfrac{\mathscr{F}(\text{IFunc}(A,B))}{\mathscr{F}_B}$, with two coefficients still indeterminate.
3. Compute the sum of the squared magnitudes of the $n-2$ known coefficients in the last step; subtract the result from $n \times$ #A to get $2r^2$ and hence r, the magnitude of the missing coefficient.
4. Compute the inverse Fourier transform of $\mathscr{F}_A$ as a function of the missing coefficient $re^{i\varphi}$, where only φ remains unknown.
5. Taking into account that all the values computed in the last step must be 0's or 1's, determine φ; complete the computation of $\mathbf{1}_A$.

To some extent, this algorithm could be used even when A is a multiset.

3.3.3 Linear programming

The matricial formalism mentioned in Section 1.2.3 provides practical solutions to many retrieval problems. In [13], we have used linear programming to good effect for solving equations like $s * \mathbf{1}_A = \mathbf{1}_B$ (which corresponds to finding a linear combination of translates of A equal to B) and the same procedure could be used for solving $\mathbf{1}_A * \mathbf{1}_{-B} = \text{IFunc}(A,B)$ in A, i.e. Lewin's problem, among others like tiling.

Here is the algorithm: given a motif A and a period n for the tiling, consider a vector $x = (x_0, \dots, x_{n-1})$. By linear programming, minimize $x_0 + x_1 + \dots x_n$ under the constraint $\mathscr{A}.x^T = (1, 1, \dots 1)$ (this is the tiling condition) and conditions $0 \leqslant x_i \leqslant 1$ for all i (this compels the 'quantity of pc i' to be somewhere between 0 and 1, and hopefully either one or the other).

But though the algorithm seems to work well, it is not formally proved yet that it always provides a solution! For one thing, there may well be multiple solutions (obtained by varying the starting point). For example, for $B = \{0, 2, 4, 6, 8, 10\} \subset \mathbb{Z}_{12}$, $\text{IFunc}(A, \pm B)$ does not change when A is replaced by $A + 2$ and there are at least six different solutions for the same value of $\text{IFunc}(A, B)$. Notice that this method bypasses Fourier transform altogether.

It is advantageous to use an environment wherein linear programming is already implemented (Mathematica, Maple, Fortran, ...).

3.3.4 Searching for Vuza canons

Tilings by translation, i.e. decomposition of cyclic groups in direct sums, gave rise to many conjectures. So far, most of them have proved to be false:

1. Sands and Tidjeman independently believed that any rhythmic canon is decon-catenable, i.e. when $A \oplus B = \mathbb{Z}_n$ then – assuming $0 \in A \cap B$ up to translation – either A or B lies in a strict subgroup of $\mathbb{Z}_n$.

2. Call D the diameter of a finite set of integers A (i.e. up to translation $A \subset \{0,1\ldots,D\}$), $\mathscr{T}$ the least period of a tiling by A (i.e. A tiles $\mathbb{Z}_\mathscr{T}$) and $\mathscr{T}(D)$ the largest $\mathscr{T}$ for all A's with diameter $\leqslant D$. From the case $A = \{0,D\}$, it is clear that $\mathscr{T}(D) \geqslant 2D$; in the other direction, from the pigeonhole principle, it can be shown that $\mathscr{T}(D) \leqslant 2^D$, a rather wide bracket.

The first conjecture was proved false by Szabó ([81]). For the second one, Kolountza-kis and others proved that $\gamma D^2 \leqslant \mathscr{T}(D) \leqslant \beta \exp(\alpha \sqrt{D \log D})$ for some constants α, β, γ; the lower bound was since increased to any power of D. The upper bound actually uses Fourier analysis, the factorisation in cyclotomic polynomials, and a so-phisticated lower bound for Euler's totient function $\varphi(n) \geqslant \dfrac{Cn}{\log\log n}$ allowing one to construct cyclotomic factors with large degrees. In this section, we will focus on the construction that proves the lower bound and on the similar one by Szabó that disproves Sand's conjecture.

Both constructions start from two basic ideas: first, for composite n, $\mathbb{Z}_n$ can be decomposed as a direct sum (or product) of other cyclic groups (three at least in both cases), enabling one to look at 3D periodic lattices; and second, a very regular tiling (say B is a subgroup of $\mathbb{Z}_n, B = d\mathbb{Z}_n$ and A is a complete set of residues modulo d) can be easily perturbed into a very aperiodic tiling. Szabó and Kolountzakis differ in the second part because their aims are different.

Generalised Kolountzakis algorithm

Initially, Kolountzakis starts from an integer $n = 30pq$ and the isomorphism $\mathbb{Z}_n \approx \mathbb{Z}_{3p} \times \mathbb{Z}_{5q} \times \mathbb{Z}_2$ where p,q are large distinct primes with a similar magnitude $\sim D$. He then singles out the two "parallel planes" $P_0 = \mathbb{Z}_{3p} \times \mathbb{Z}_{5q} \times \{0\}$ and $P_1 = \mathbb{Z}_{3p} \times \mathbb{Z}_{5q} \times \{1\}$. He starts from the trivial tiling of $\mathbb{Z}_{3p} \times \mathbb{Z}_{5q}$ by $A = \{0,1,2\} \times \{0,1,2,3,4\}$ and $B = \{0,3,6\ldots\} \times \{0,5,10\ldots\}$ where B is a subgroup (isomorphic to $\mathbb{Z}_{pq}$) and A, omitted in Fig. 3.15, would appear as a small square. Now for P_0, *a row of* the first factor of B is translated; say $B_0 = \{0,4,6\ldots\} \times \{0,5,10\ldots\}$ and similarly for P_1 we translate *a column* of the second factor, say $B_1 = \{0,3,6\ldots\} \times \{0,8,10\ldots\}$. This shatters any periodicity in the tiling. Keeping the same $A \times \{0\}$ as motif and putting $B' = B_0 \times \{0\} \cup B_1 \times \{1\}$ we have an aperiodic tiling of $\mathbb{Z}_{3p} \times \mathbb{Z}_{5q} \times \mathbb{Z}_2$ and by isomorphism a Vuza canon in $\mathbb{Z}_n$. Explicit expression of this isomorphism (given below in 3.3) shows that the diameter of A has the same order of magnitude as $D \sim p \sim q$, whilst $n = 30pq \sim D^2$, proving the worst-case lower bound given *supra*. In Fig. 3.15 we can see at left the regular lattice B, and at right the same perturbed in B'; the first and second planes having respectively a row and a column pushed somewhat out of place (p, q have been reduced to 3 and 4 for the sake of readability).

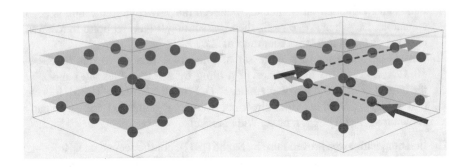

Fig. 3.15. A lattice tiling and its perturbation

I provide without details an example of such a construction, which is useful for building Vuza canons of medium size even though it was devised to prove asymptotic results.

Let $p = 3, q = 5$. Hence $n = 450$. In $\mathbb{Z}_n$ we find that A is $\{0, 126, 252\} \oplus \{0, 100, 200, 300, 400\}$, i.e.

$$A = \{0, 2, 28, 54, 100, 126, 128, 154, 200, 226, 252, 254, 326, 352, 378\}$$

and $B = \{0, 30, 60, 90 \ldots 420\} = 30\,\mathbb{Z}_n$. This corresponds, in 3D, to the triplets with coordinates $(0/3/6, 0/5/10/15/20, 0/1)$ (/ denotes here an arbitrary choice between the values). The perturbation changes $(0, 0, 0)$ and $(3, 0, 0)$ to $(2, 0, 0)$ and $(5, 0, 0)$ in B_0, and the $(3, 5k, 1)$ to $(5, 5k + 2, 1)$ in B_1, which yields ultimately the new factor

$$B' = \{15, 21, 30, 45, 60, 90, 100, 105, 111, 120, 135, 180, 195, 201, 210, 225,$$
$$240, 250, 270, 285, 291, 315, 330, 360, 375, 381, 390, 400, 405, 420\}.$$

By using five parallel planes instead of two, it is possible to get a tiling of $\mathbb{Z}_{180}$, the minimal value for this construction. One solution is shown in Fig. 3.16.

I will now expound a more general version.

1. Have five numbers a, b, c, p, q such that ap, bq and c are pairwise coprime.
2. Construct the tile
   $A \subset G = \mathbb{Z}_{ap} \times \mathbb{Z}_{bq} \times \mathbb{Z}_c$ by $A = \{0, 1, 2 \ldots a - 1\} \times \{0, 1 \ldots b - 1\} \times \{0\}$.
3. Construct the lattice complements

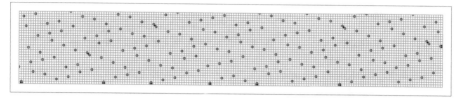

Fig. 3.16. Minimal Vuza canon ($n = 180$) built by Kolountzakis' algorithm

$$B_0 = \{0, a, 2a, \ldots (p-1)a\} \times \{0, b \ldots (q-1)b\} \times \{0\} \ldots$$

$$\vdots$$

$$B_{c-1} = \{0, a, 2a, \ldots (p-1)a\} \times \{0, b \ldots (q-1)b\} \times \{c-1\}.$$

4. For $k = 0 \ldots c-1$ add a perturbation vector ε_k to every element of each B_k, either of the form $\varepsilon_k = (p_k, 0, 0)$ or $\varepsilon_k = (0, p_k, 0)$. The two kinds must be present. Let $B'_k = B_k + \{\varepsilon_k\}$.
5. Compute $B = \bigcup B'_k$. Now $A \oplus B = G$.
6. Turn into a tiling of $\mathbb{Z}_n$ by the canonical linear isomorphism $\Psi : G \to \mathbb{Z}_n$,

$$\Psi(x, y, z) = ux + vy + wz$$

where u is defined modulo by $\Psi(x, y, z) \equiv x \pmod{a}p$ and similar equations,

hence $\begin{cases} u \equiv 1 \pmod{a}p \\ u \equiv 0 \pmod{b}q \\ u \equiv 0 \pmod{c} \end{cases}$; so u is a multiple of bcq and we get explicitly

$u = bcq \times (bcq)^{-1}$ in $\mathbb{Z}_{ap}$ (similarly for v, w).

This is not guaranteed to yield a Vuza canon, though it usually does. In practice, generate all possible canons by this method and sort out the aperiodic ones.

On the other hand, it is possible to compute R_A quite easily; since in $\mathbb{Z}_n$ one gets

$$A = \{0, u, \ldots (a-1)u\} \oplus \{0, v, \ldots (b-1)u\}$$

and in polynomials

$$\mathbf{A}(X) = (1 + X^u + X^{2u} + \ldots X^{(a-1)u})(1 + X^v + X^{2v} + \ldots X^{(b-1)v}) = \frac{X^{au} - 1}{X^u - 1} \frac{X^{bv} - 1}{X^v - 1}.$$

Hence R_A is made of the divisors of au which do not divide u, together with the divisors of bv which do not divide v:

Proposition 3.62. $R_A = \big(\mathrm{Div}(au) \cup \mathrm{Div}(bv)\big) \setminus \big(\mathrm{Div}(u) \cup \mathrm{Div}(v)\big)$.

This easily entails the non-periodicity of A. It is also a clear case of verifying conditions (T_1) and (T_2). It is possible to tell something about R_B (notably proving that it always satisfies condition (T_2)), but since the computation is analogous in the next algorithm, I will only do the latter.

Szabó's algorithm

In [81] the 3D-decomposition is not explicitly made. I will endeavour here to make it so.

Consider three pairs of integers $u_i, v_i, i = 1 \ldots 3$ such that $u_i v_i$ and $u_j v_j$ are coprime for $i \neq j$. Let $m_i = u_i v_i$ and $n = m_1 m_2 m_3$. It is convenient to introduce $g_i = n/m_i$, e.g. $g_1 = u_2 v_2 u_3 v_3$.

For an example, let $u_1 = v_1 = 2, u_2 = v_2 = 3, u_3 = v_3 = 5, n = 900$.

Now the three groups G_i generated by the m_i satisfy $G_1 \oplus G_2 \oplus G_3 = \mathbb{Z}_n$. Each can be further decomposed in

$$G_i = \{0, g_i, 2g_i, \ldots (u_i - 1)g_i\} \oplus \{0, \frac{n}{v_i}, 2\frac{n}{v_i}, \ldots (v_i - 1)\frac{n}{v_i}\} = A_i \oplus B_i.$$

In the example, $G_2 = \{0, 100, 200\} \oplus \{0, 300, 600\}$.

Construct $A = \bigoplus A_i, B = \bigoplus B_i$: we have a tiling since

$$\mathbb{Z}_n = \bigoplus_i (A_i \oplus B_i) = \left(\bigoplus_i A_i\right) \oplus \left(\bigoplus_i B_i\right) = A \oplus B.$$

It is helpful to think of A as 'small change' and B as 'banknotes'.[38]

In the example, $A = \{0, 225\} \oplus \{0, 100, 200\} \oplus \{0, 36, 72, 108, 144\}$ and $B = \{0, 30, 60 \ldots\} = 30\mathbb{Z}_{900}$.

B is always a subgroup, generated by all three $n/v_i = u_i g_i$, i.e. $B = \frac{n}{v_1 v_2 v_3} \mathbb{Z}_n$.

The idea is to perturbate B using the three dimensions. To ensure that the new B' still tiles with A, Szabó chooses a (circular) permutation σ of $\{1, 2, 3\}$. Remembering that the elements of B can be written as $\sum k_i u_i g_i$, select the $x_{k,i} = k u_i g_i + u_{\sigma(i)} g_{\sigma(i)}$ and replace all $x_{k,i}$ by $x'_{k,i} = x_{k,i} + g_i$.

In the example, if we take $\sigma(i) = i + 1 \pmod 3$ then we replace $x_{2,3} = 2u_3 g_3 + u_1 g_1 = 360 + 450 = 810$ by $x_{2,3} + g_1 = 810 + 225 = 135$. On this term, the divisibility by 2 is destroyed, this is how this construction shatters Sand's conjecture. In all, $\sum v_i = 2 + 3 + 5 = 10$ elements are changed.

This destroys the regularity of B but preserves the tiling quality, and perhaps a little more:

Theorem 3.63. *This construction yields a non-deconcatenable, non-demultiplexable, Vuza canon for large enough u_i, v_i. However, both factors of the tiling always satisfy condition (T_2).*

The first assertion is proved in [81], at least for composite n greater than 60,060 (though the smallest known counterexample, which uses this construction, lies in $\mathbb{Z}_{900}$). The last assertion appears in the literature, but as far as I know no proof of it has been published before.

---

[38] Appropriately, one of the very first papers on tilings of integers, *On Number Systems* by Nicolas de Bruijn (1956), originated from the consideration of the British money system.

Proof. Consider the characteristic polynomial

$$\mathbf{B}(X) = 1 + X^m + X^{2m} + \cdots = \frac{X^n - 1}{X^m - 1} \quad \text{where} \quad m = u_1 u_2 u_3.$$

In order to turn B into B' we multiply, for all i and all $k = 0 \dots v_i - 1$, the term $X^{ku_i g_i + u_{\sigma(i)} g_{\sigma(i)}}$ by X^{g_i}. In effect we had to $\mathbf{B}(X)$ the polynomials $P_i(X), i = 1 \dots 3$ defined by

$$(X^{g_i} - 1) X^{u_{\sigma(i)} g_{\sigma(i)}} \sum_{k=0}^{v_i - 1} X^{ku_i g_i} = (X^{g_i} - 1) X^{u_{\sigma(i)} g_{\sigma(i)}} \frac{X^n - 1}{X^{u_i g_i} - 1}.$$

Adding these polynomials and multiplying by the explicit form of $\mathbf{A}(X)$ *would prove that the new outer voice* B' *still tiles with* A*. I will not do it here, since it is already done in [81].*

The cyclotomic factors of this perturbation factor are the Φ_d with $R_{P_i} = d \in (\text{Div}(n) \setminus \text{Div}(u_i g_i)) \cup \text{Div}(g_i)$. Remember that $R_B = \text{Div}(n) \setminus \text{Div}(m)$. Let us elucidate S_B: a prime factor p of u_i can only appear again in v_i by assumption; if it does not then it is cancelled out in the divisors of m, i.e. the prime powers in S_B are those common to u_i and v_i. Any such prime factor being confined to one index i can be labelled p_i, and $p_i^k \in S_B$ only if k is greater than the p_i-valuation of u_i, i.e. p_i^k is *not* a divisor of u_i.

In the example above, $\mathbf{B}(X) = \dfrac{X^{900} - 1}{X^{30} - 1}$ *and* $S_B = \{2 \times 2, 3 \times 3, 5 \times 5\}$.

Such powers still belong to R_{P_i}. So do products of these powers for different indexes i: consider without loss of generality $r = p_1^2 p_2^2$ where p_i is a prime factor of u_i and $v_i, i = 1, 2$ (with valuation 1 to ease the notation). Then r is a divisor of n, of course, but not a divisor of $u_1 g_1 = n/v_1 = p_1^1 \times Q$ where Q is coprime with p_1. A similar verification can be done for P_3. This means that condition (T_2) still holds.

In the example above, $S_{B'} = S_B = 4, 9, 25$ *and we preserve at least* $36, 100, 225$ *and* 900 *in* $S_{B'}$*. Some factors have disappeared but are not required by condition* $(T_2) : 12, 18, 20, 45, 50, 60, 75, 90, 150, 180, 300, 450.$

Matolcsi's algorithm

In [57], Matos Matolcsi devised a neat procedure for an exhaustive search for Vuza canons in a given $\mathbb{Z}_n$. Though this sometimes fails because of computational complexity, it is still worthwhile to study it in the context of this book.

The key to his procedure is a useful lemma in [35]:

Lemma 3.64. *If A satisfies (T_1) and (T_2), then a complement of A in $\mathbb{Z}_n$, i.e. B satisfying $A \oplus B = \mathbb{Z}_n$, can be produced by its characteristic polynomial: $B(X)$ is the product of the $\Phi_{p^\alpha}(X^{n/p^{v(p)}})$, where $p^\alpha \mid n$ is not in S_A, and $n = \prod_i p_i^{v(p_i)}$ is the decomposition of n into prime powers (so that $n/p^{v(p)}$ is the largest divisor of n coprime with p).*

Example 3.65. Consider $S_A = \{2, 8\}$ and $n = 24$.[39] Since $24 = 2^3 3^1$, the missing prime powers in S_A – which must indeed be in S_B – are $4 = 2^2$ and 3, which are respectively complemented to 24 by coprime prime powers 3 and 8. We compute

$$B(X) = \Phi_4(X^3) \times \Phi_3(X^8) = (1 + (X^3)^2)(1 + (X^8) + (X^8)^2),$$

hence $B = \{0, 6, 8, 14, 16, 22\}$ which does tile, for instance with $A = \{0, 3, 12, 15\}$.

Now the idea is to check all possible sets S_A. Begin by choosing n.

- Compute all partitions in two subsets of the set of prime power divisors of n. Keep (usually) the smallest part, which will be S_A (the other being of course S_B).
- Compute the Coven-Meyerowitz complement B for S_A.[40]
- Compute all possible A completing B, using one of the general completion algorithms described in [11].[41] Sort by the different values of R_A, keeping one representative A_i for each possibility.
- Discard all sets R_A that either
  1. ensure that A is periodic, or
  2. ensure that B must be periodic (recalling that R_B must contain at least all divisors of n not in R_A), making use of Theorem 3.28.
- For each remaining representative of possible A's, compute complements B, discarding eventual periodic ones.
- Whatever remains is a Vuza canon.

Details and tables of results are given in [8].

One algorithm that I will not discuss here, though it sounds closely related to harmonic analysis, is the search for a spectrum (cf. Section 3.2.6). Actually it is mostly (as of today) a computational problem; Kolountzakis has studied its complexity and provides strong heuristic reasons for it to be NP-complete, unless the (T_2) conjecture is true. Actually he views this as a strong argument against the latter conjecture!

Exercises

Exercise 3.66. Compute the cyclotomic polynomial Φ_d when d runs over all divisors of 12 (use Eq. 3.1).

Exercise 3.67. $X^8 + 1$ is a cyclotomic polynomial. Which one?

Exercise 3.68. Choose some singular pc-set in Table 8.2 and check which of Lewin's conditions is satisfied. Compare with the appropriate Fourier coefficient (e.g. if the augmented triad property is satisfied, check that $a_3 = 0$).

---

[39] If we start from an actual motif A and n is unknown, n can be taken equal to the lcm of R_A – or any multiple thereof.

[40] This is a simple motif, product of 'metronomes', cf. exercises.

[41] This is the weak point of the algorithm because when B is very regular, both the number of solutions for A and the searching time get considerable.

Exercise 3.69. Is $\{0,2,3,5,7,8\}$ singular or invertible in $\mathbb{Z}_{12}$?

Exercise 3.70. Express a fifth (e.g. $\{0,7\}$) as a linear combination of the 11 other ones.

Exercise 3.71. Compute by hand the DFT of $\{0,1,6,7,11\}$, Berg's sonata's initial pc-set.

Exercise 3.72. Decompose the even elements of $\mathbb{Z}_{32}$ in classes of associated elements, i.e. according to their order.

Exercise 3.73. $A = \{0,1,7,11,17,18,24\} \subset \mathbb{Z}_{30}$. Check that $a_1 = 0$ and that A cannot be decomposed as a reunion of regular polygons.

Exercise 3.74. Prove Proposition 3.20 and/or the next one.

Exercise 3.75. Check that $A = \{0,1,6,10,12,13,15,19\}$, $A' = \{0,2,5,6,11,12,15,17\}$ both tile $\mathbb{Z}_{24}$.

Exercise 3.76. Compute R_A for $A = \{0,5,8,13\}$.

Exercise 3.77. Use Jedrzejewski's recipe and build a Vuza canon.

Exercise 3.78. Prove Theorem 3.49 (discuss on each possible residue i, or read [6]).

Exercise 3.79. Check that $\{0, \dfrac{1}{8}, \dfrac{1}{2}, \dfrac{5}{8}\}$ is a spectrum for $A = \{0,1,4,5\}$ in $\mathbb{Z}_8$.

Exercise 3.80. Finish the computation of the example in $\mathbb{Z}_{900}$ of Szabó's algorithm.

Exercise 3.81. A motif A is such that $S_A = \{2,8,9\}$ and satisfies condition (T_2). Build B that tiles with A using the construction in Lemma 3.64. Use

$$\Phi_{p^\alpha}(X) = 1 + X^{p^{\alpha-1}} + X^{2p^{\alpha-1}} + \ldots X^{(p-1)p^{\alpha-1}} = \frac{X^{p^\alpha} - 1}{X^{p^{\alpha-1}} - 1}.$$

4

Saliency

Summary. In the seminal [72], Ian Quinn tries to define a 'landscape of chords' starting from cultural/intuitive knowledge of the most 'salient' chords, and from there infers in a prodigious leap of intuition the existence of a measurable 'chord quality', or saliency, maximal for the prototypical chords. Moreover, he notices that these chords are well known: they are the Maximally Even Sets, i.e. the most even divisions of the octave. In another brilliant intuition, he notices that such pc-sets are characterised by a maximal value of some Fourier coefficient. Thus his vision of a chord landscape is achieved by plotting the magnitude of this Fourier coefficient for all chords (with a given cardinality). Though other measures of chord quality have been devised (Douthett-Kranz, Junod), this notion of saliency will of course be the topic of this chapter.

It is important to mention that this notion applies equally well to periodic rhythms, or any (musical) phenomenon that can be modeled in a cyclic group; for instance, the *tresilo* which is prominent in much of Latin-American dance music will be mentioned below. But since the focus in correlated research has been on scales, I will stick mostly to pc-sets vocabulary and examples.

A selection of Fourier profiles (i.e. magnitudes of Fourier coefficients) of pc-sets is shown in Chapter 8. In this chapter, many references are made to these pictures and the reader is invited to browse the whole collection online at

http://canonsrythmiques.free.fr/MaRecherche/photos-2/

(pc-sets are considered up to transposition but not inversion for easier recognition).

Alternatively, the reader is invited to download some software for computing their own Fourier coefficients of any pc-set on

http://canonsrythmiques.free.fr/MaRecherche/styled/.

This requires Mathematica[TM] or the free CDF reader provided by Wolfram Research.

We will study three types of pc-sets with some overlapping between them: saturated scales, generated scales, and maximally even scales. All these highly polarised sets of notes have highly uneven magnitudes of Fourier coefficients; actually, all of them are characterised by some maximum Fourier coefficient. Once this classification is achieved, and some similar/close cases examined, we can move on to the opposite case, flat histogram of either intervals or magnitudes of Fourier coefficients, and prove that the one is flat if and only if the other is too. A seminal case of a flat profile is the aggregate minus one note, which is indeed often tiled by such subsets. Thus the landscape of chords/scales is well described by its peaks and valleys. For instance, the highest peaks in Fig. 4.1 for trichords are augmented triads.

© Springer International Publishing Switzerland 2016

E. Amiot, *Music Through Fourier Space*, Computational Music Science,
DOI 10.1007/978-3-319-45581-5_4

Fig. 4.1. The landscape of trichords

4.1 Generated scales

Much study has been devoted in music theory to the generation of musical scales, whether with just intervals (fifths, thirds) or otherwise. In this section we will consider the monogenous case in equal temperament, according to the following:

Definition 4.1. *A generated scale in* $\mathbb{Z}_n$ *is a subset*[1] *of* $\mathbb{Z}_n$ *generated by some arithmetic progression, i.e.* $A = \{a, a+f, a+2f, \ldots a+(d-1)f\}$. *The generating interval*[2], *or generator, or common difference, is* f, *the starting point is* a.

The most famous example is the diatonic scale, generated by fifths (or fourths). Other cases are the non-hemitonic pentatonic ('Chinese') scale and the whole-tone scale. These three are maximally even scales (see Section 4.2), which is not the case of the Guidonian hexachord $\{0, 2, 4, 5, 7, 9\}$ though it is also generated.

---

[1] We require distinct elements, i.e. A is not a multiset. Of course A can be viewed as a periodic rhythm instead of a scale, but the historical context of study of these subsets being scale theory, the name stuck.

[2] The letter f is chosen as the initial of 'fifth', but of course it can take on any value.

4.1.1 Saturation in one interval

Since $a + kf$ can only be connected by an interval of f to $a + (k+1)f$ (upwards) or $a + (k-1)f$ (downwards), the number of occurrences of one given interval in a pc-set cannot exceed the set's cardinality. Conversely, we get the saturation characterisation:

Proposition 4.2. *If a scale A with d elements is generated by interval f, then the number of occurrences of f is $d-1$ or d. The latter case is that of a closed regular polygon. Conversely, a saturated scale is, in the latter case, a periodic subset or a reunion of periodic subsets with the same size (i.e. the orbit of a subgroup of $\mathbb{Z}_n$); and in the former, the same but with one incomplete subcycle.*

The more complicated case of several complete plus one incomplete cycles occurs fairly frequently in 19th century music, cf. the excerpt of Liszt's Piano Sonata in Fig. 4.2 featuring $\{2,5,8,11\} \cup 9$ and $\{1,4,7,10\} \cup 11$. Its Fourier profile appears in Fig. 8.21.

Fig. 4.2. Minor third with multiplicity 4 in 5 notes, in Liszt's Sonata in B.

We will find similar subsets when computing the maximal possible values of the magnitude of Fourier coefficients.

Proof. The number of occurrences of f in $\{a, a+f, a+2f, \ldots a + (d-1)f\}$ is clearly at least $d-1$ and can only reach d if $a + df = a$ (in $\mathbb{Z}_n$), which means that $df = 0 \mod n$; and hence the scale closes, i.e. A is a regular polygon. Conversely, the pairs $(x, x+f)$ cannot happen more than d times, in which case every single element $x \in A$ plays once the role of x in the pair and once the role of $x+f$, i.e. one has $x + f \in A$ and $x - f \in A$ (equivalently, the map $\tau_f : a \mapsto a + f$ is a permutation of the set A). This means that A is closed under translation by f, i.e. A is an orbit, or a reunion of orbits, of the group $f\mathbb{Z}_n$, i.e. a reunion of translates of $f\mathbb{Z}_n$. With a count of $d-1$ occurrences of interval f, the condition can and must be relaxed on one and only one x, which will satisfy x and $x - f \in A$ but $x + f \notin A$, so that by removing that element we get the same case with both #A and the number of occurrences of f decremented by one; so the proposition is proved by induction.

4.1.2 DFT of a generated scale

It is easy to compute the DFT of chromatic cluster $A = \{0,1,2,\ldots d-1\}$, since all coefficients are sums of geometric series:

$$\mathscr{F}_A(t) = \sum_{k=0}^{d-1} e^{-\frac{2ikt\pi}{n}} = \frac{e^{-\frac{2idt\pi}{n}}-1}{e^{-\frac{2it\pi}{n}}-1} = \frac{e^{-\frac{idt\pi}{n}}}{e^{-\frac{it\pi}{n}}}\frac{e^{\frac{idt\pi}{n}}-e^{-\frac{idt\pi}{n}}}{e^{\frac{it\pi}{n}}-e^{-\frac{it\pi}{n}}} = e^{i(1-d)t\pi/n}\frac{\sin\frac{dt\pi}{n}}{\sin\frac{t\pi}{n}}.$$

Hence the magnitude of the DFT of any generated scale

$$B = fA + \tau = \{\tau, \tau+f, \tau+2f, \ldots\}$$

(translation by τ does not change the magnitude, and multiplication by f multiplies the index of the coefficient):

Proposition 4.3. $|\mathscr{F}_B(t)| = \begin{cases} d & \text{if } \sin\frac{f\pi t}{n} = 0 \ (i.e. \ n \mid ft) \\ (\pm)\dfrac{\sin\frac{fd\pi t}{n}}{\sin\frac{f\pi t}{n}} & else \end{cases}.$

For instance the value of $|\mathscr{F}_A(5)|$ when A is a diatonic scale is

$$-\frac{\sin\frac{5\times7\pi5}{12}}{\sin\frac{7\pi5}{12}} = \frac{\sin\frac{7\pi}{12}}{\sin\frac{\pi}{12}} = \frac{1}{\tan\frac{\pi}{12}} = 2+\sqrt{3}.$$

It is obvious that the first case, d, is the maximum possible value, especially when one remembers that we just summed d complex numbers $e^{-\frac{2ikft\pi}{n}}$, all of them with magnitude 1. It is perhaps less obvious that the reciprocal is true (for the moment, we consider only generated scales): if any of the exponentials in the sum defining the Fourier coefficient do not have the exact same direction, then their sum has a smaller length than the sum of their lengths:

Lemma 4.4. *For $a,b \in \mathbb{C}$, $|a+b| = |a| + |b| \iff a,b$ have the same direction, i.e. $\exists \lambda \in \mathbb{R}_+, b = \lambda a$ (unless $a = 0$).*

So when the magnitude of the Fourier coefficient is maximum, all exponentials in it share the same direction. But equality of the phases of all $e^{-2ifkt\pi/n}$ means that $n \mid ft$, i.e. we are in the first case when $\sin\frac{f\pi t}{n} = 0$.

The other extreme case is $\mathscr{F}_B(t) = 0$, when bdt is a multiple of n but bt is not. Let us clarify the behavior of these values. Jason Yust noticed the periodicity of these coefficients:

Proposition 4.5. *Fix the generator f and the index of the Fourier coefficient, t. Then the magnitude[3] of this Fourier coefficient is periodic in the cardinality d of the generated scale: $d \mapsto |\mathscr{F}_B(t)|$ has period $\dfrac{n}{\gcd(n, ft)}$.*

For $n = 12$, this period boils down to:

---

[3] The complex Fourier coefficient itself is either periodic or anti-periodic.

- n/r, where r is the integer closest to 0 and congruent to $\pm ft$; and
- no period (i.e. period 12) when ft is coprime with 12 (for instance, $\mathscr{F}_B(1)$ for fifth-generated scales has no period).

A few examples will show how simple this is:

Example 4.6. Consider first chromatic clusters, like $\{0,1,2\}$, with generator 1 and let us look at $\mathscr{F}_B(4)$ as a function of the cardinality d: $|\mathscr{F}_B(4)| = \left| \dfrac{\sin(d\pi/3)}{\sin(\pi/3)} \right| = \psi(d)$ and ψ is 3-periodic ($|\sin|$ being π-periodic). Indeed the values taken for $d = 1,2,3\ldots$ are $1,1,0,1,1,0,1,1,0,1,1\ldots$.

For a less trivial case, take coefficient 5 and generator 2 (whole-tone scale chunks). Since $2 \times 5 = 10 = -2 \mod 12$ we have $r = 2$, period 6, and indeed for $d = 1,2,3\ldots11$ we compute $|\mathscr{F}_B(5)| = 1, \sqrt{3}, 2, \sqrt{3}, 1, 0, 1, \sqrt{3}, 2, \sqrt{3}, 1$. The associated pc-sets appear in the tables as Figs. 8.4, 8.8, 8.15, 8.20, and 8.23.

A more complicated case where Yust's rule of thumb does not apply: let $n = 24$ and $f \times t = 7 \times 2 = 14$. Then the period is 12.

Lastly, a rhythm example: consider generator 3 in an eight beats bar; the *tresilo* $(0, 3, 6)$ (modulo 8) is such a generated rhythm, with $d = 3$. The value of the Fourier coefficient $|a_3|$ takes on magnitudes $\frac{\sin 9d\pi/8}{\sin 9\pi/8}$, which is maximum when $d = 4$ for rhythm $(0, 1, 3, 6)$. In general, $d \mapsto \frac{\sin \frac{fd\pi t}{n}}{\sin \frac{f\pi t}{n}}$ will be maximum when fdt is as close as possible to $n/2 \mod n$.

The proof of this periodicity lies in the formula in Proposition 4.3. Amusingly, Yust's shortcut for $n = 12$ works for the same reason that Lemma 4.20 below is true.

Another beautiful relationship between the chromatic case (generator 1) and the general case (generator f) is

Theorem 4.7 (P. Beauguitte, 2011). *Let $A_k = \{0,1,2\ldots k-1\} \subset \mathbb{Z}_n$. For k coprime with n, let $\ell = k^{-1}$ be the multiplicative inverse of k modulo n and $B = -kA_\ell = \{0, -k, -2k \cdots -k(\ell-1)\}$ the ℓ-scale generated by $-k$. Then $\mathscr{F}_B = 1/\overline{\mathscr{F}_A}$, i.e. the coefficients of one scale are the inverses of the coefficients of the other.*[4]

The choice of ℓ will be clarified below with the definition of ME sets. A common example with $k = 7, n = 12, \ell = 7$ yields the diatonic scale, but in general, the two scales have a different number of elements.

Proof. $\mathscr{F}_A(t) = 1 + e^{-2i\pi t/n} + \ldots e^{-2i\pi(k-1)t/n} = \dfrac{1 - e^{-2i\pi kt/n}}{1 - e^{-2i\pi t/n}}$ and

$\mathscr{F}_B(t) = 1 + e^{2i\pi kt/n} + \ldots e^{2i\pi(\ell-1)kt/n} = \dfrac{1 - e^{2i\pi k\ell t/n}}{1 - e^{2i\pi kt/n}} = \dfrac{1 - e^{2i\pi t/n}}{1 - e^{2i\pi kt/n}}$, hence the result by inverting the fraction and the phases.

This remarkable result shows that for many generated scales, the direction of the DFT is the same as for a chromatic sequence, whilst the magnitude is inversed. This appears clearly in Fig. 4.3, with $n = 10, k = 3, \ell = 7$:

---

[4] Except of course for index 0 which is the cardinality of the scale.

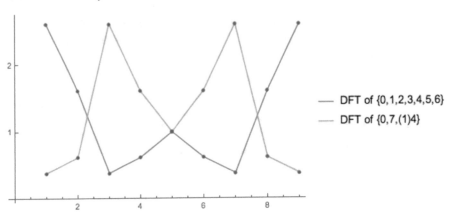

$$- \text{DFT of } \{0,1,2,3,4,5,6\}$$
$$- \text{DFT of } \{0,7,(1)4\}$$

Fig. 4.3. Beauguitte's theorem: inverse magnitudes of two generated scales in $\mathbb{Z}_{10}$.

The saturation feature is linked with the *probability of occurrence* of intervals: in diatonic music, the fifth is more probable than other intervals (if the probability of any pitch-class is uniform, which admittedly is seldom the case except perhaps in strict dodecaphonic, non-serial music), as checked experimentally in [58] for instance. This suggests, in a broad sense, that generated scales are somewhat *periodic* and might be recognised by Fourier features. This is precisely the topic of the maximally even sets section below. For more about occurrences of intervals and their relationship with Fourier coefficients, see Section 4.3.

4.1.3 Alternative generators

Notice the extreme cases (first pointed out, to the best of my knowledge, by N. Carey in [28] wherein the first case of Theorem 4.8 is also proved) when f is a generator of $\mathbb{Z}_n$, and A is the whole aggregate, or $d = n - 1$, i.e. A is the whole group $\mathbb{Z}_n$ minus one element. In this case, A has $\varphi(n)$ distinct generators[5] (and as many starting points), which is a somewhat unexpected behaviour for arithmetic sequences. For instance, the aggregate from C to $B\flat$, e.g. $\{0, 1, 2, 3 \ldots 10\}$, can be written as four distinct arithmetic sequences:

$$(0, 1, 2 \ldots 10), (4, 9, 2, 7, 0, 5, 10, 3, 8, 1, 6) \text{ and their reverses, with generators } 11, 7.$$

This can be seen in Fig. 4.4 with 6 different generators for a 7-scale in $\mathbb{Z}_{21}$.
    The converse is true:

Theorem 4.8. *[Amiot, 2011] The number of generators of a generated scale is always a totient number, i.e. $\varphi(n)$ for some n.*
    More precisely:

---

[5] Remember φ is Euler's totient function, which gives precisely the number of generators of a cyclic group.

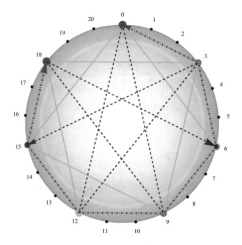

Fig. 4.4. Many generators for a regular polygon

⬦ *if f is coprime with n then A has exactly two generators ±f, unless A is the full aggregate (A = $\mathbb{Z}_n$) or the almost full ($\mathbb{Z}_n \setminus \{u\}$).*
⬦ *if f is not coprime with n, the generated scale A with cardinality d > 1, has*

- *one generator when the scale is (a translate of) $\{0, n/2\}$ (a tritone);*
- *two generators (not coprime with n) when d is strictly between 1 and $n' - 1 = (n/m) - 1$ where $m = \gcd(n, f)$;*
- *$\varphi(d)$ generators when $d = n' = n/m$, i.e. when A is a regular polygon;*
- *$\varphi(d+1)$ generators when $d = n' - 1$, A is a regular polygon minus one vertex.*

The last two cases are those of a full or almost full regular polygon, whose picture is the same as the full or almost full aggregate but for a smaller cardinality $n' \mid n$. Moreover all generators share the same order in the group $(\mathbb{Z}_n, +)$.

Proof. First consider the case of a generator f coprime with n. Up to multiplication by the inverse f^{-1} of this generator modulo n and translation, we are dealing with the chromatic sequence $A = \{0, 1, \ldots d - 1\}$ and we are looking for an alternative generation to the obvious one (generator 1). So let us assume that A can also be generated as $A = \tau + b \times \{0, 1, 2 \ldots d - 1\} = bA + \tau$ and let us prove that $b = \pm 1$. My original proof made use of the interval vector of A, which is $(d, d-1, d-2 \ldots d-2, d-1)$. An alternative one, more appropriate in the context of this book, uses the DFT:[6]

---

[6] Incredibly but appropriately, a recent formula [78] expresses the totient function as the DFT of the GCD: $\varphi(n) = \sum_{k=1}^{n} e^{\frac{2i\pi k}{n}} \gcd(n, k) = \sum_{k=1}^{n} \cos\left(\frac{2\pi k}{n}\right) \gcd(n, k)$.

$$\mathcal{F}_A(t) = \sum_{k=0}^{d-1} e^{-2ikt\pi/n} = \frac{e^{-2idt\pi/n} - 1}{e^{-2it\pi/n} - 1},$$

$$\mathcal{F}_{bA+\tau}(t) = \sum_{k=0}^{d-1} e^{-2i(bk+\tau)t\pi/n} = e^{-2i\tau t\pi/n} \frac{e^{-2ibdt\pi/n} - 1}{e^{-2ibt\pi/n} - 1}.$$

It is sufficient to focus on the magnitudes: since $|e^{-2i\varphi} - 1|$ is equal to $|2\sin\varphi|$, the respective magnitudes are

$$\frac{\sin(d\pi/n)}{\sin(\pi/n)} \quad \text{and} \quad \frac{\sin(bd\pi/n)}{\sin(b\pi/n)} \qquad (0 < d < n).$$

(I removed the absolute values for readability). Replacing b if necessary by $n - b$ without changing the magnitude, one may assume without loss of generality that $b \in \{0, 1 \ldots n/2\}$. A cursory study of next-to-maximum values[7] of function $f : b \mapsto \frac{\sin(db\pi/n)}{\sin(b\pi/n)}, 1 \leqslant b \leqslant n/2$ (see Fig. 4.5) proves that b must be equal to 1 for the respective magnitudes to coincide, hence $b = \pm a$. Let us now consider f non co-

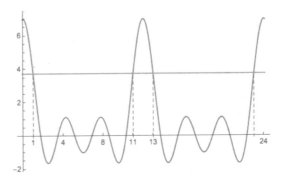

Fig. 4.5. Graph of $f : b \mapsto \dfrac{\sin(7b\pi/12)}{\sin(b\pi/12)}$

prime with n, i.e. $m = \gcd(n, f) > 1$. The cardinality of A is now less than n/m, since $\frac{n}{m}f = 0 \mod n$. The difficult question is: do we reach the same m if we start from another generator? But with a computation similar to the one above, if A is generated by f then

$$|\mathcal{F}_A(t)| = \begin{cases} \dfrac{|\sin(\pi dt f/n)|}{|\sin(\pi t f/n)|} & \text{or} \\ d & \text{when } \sin(\pi t f/n) = 0. \end{cases}$$

Moreover, $|\mathcal{F}_A(t)| \leqslant d$, and $|\mathcal{F}_A(t)| = d \iff \sin(\pi t f/n) = 0$. This entails the following:

---

[7] They occur for $b > \frac{2n}{d}$ and hence $f(x)$ does not exceed $\frac{1}{\sin(2\pi/d)}$, well under $\frac{\sin(d\pi/n)}{\sin(\pi/n)} = f(1)$.

Lemma 4.9. *If f, g are two generators of a same scale A, then*

$$m = \gcd(n, f) = \gcd(n, g).$$

NB: this lemma can also be reached algebraically, by considering the *group of differences*[8]

$$\Delta^{\infty}(A) = \lim_{n \to \infty} \Delta^n(A) = \bigcup_{n \geqslant 1} \Delta^n(A) \text{ where } \Delta(X) = X - X = \{x - y, (x, y) \in X^2\}.$$

This shorter but more abstract proof was used in [9].

Now the end is easy: up to translation, assume A contains 0. Then $A = mA'$ where the elements of A' are defined modulo $n' = n/m$, and we are back to the initial case $\gcd(n', f) = 1$ when we have only two generators, except if A' is an (almost) full aggregate. This yields the theorem.

Leaving aside the extreme cases of one-note scales and tritones, the geometry of generated scales comes in three types:

- The seminal case: 'diatonic-like scales', i.e. scales with only two (opposite) generators.
- Regular polygons.
- Regular polygons minus one note.

So this seminal case, with one beginning and one end, is by no means the only one. The three cases are summarised in Fig. 4.6.

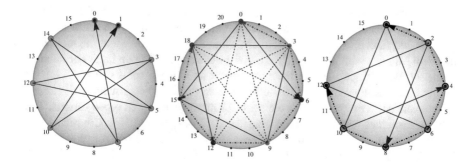

Fig. 4.6. The three cases: seminal, polygon and almost-whole polygon

4.2 Maximal evenness

Maximally even sets, or ME sets for short, were introduced in [31, 30] and developed by Jack Douthett and other co-authors. In the context of this book, his most

[8] [65], Section 7.26.

interesting paper is [33] wherein a ME set is described and defined as an equilibrium position for (say) electrons placed on several equally disposed sites on a circle; it is impressive that seven electrons on 12 sites will choose to settle as a diatonic scale!

There are many possible definitions of maximal evenness, an intuitive notion but a tricky one to nail down: see [31, 32, 38, 24]. The most practical appears in the seminal [30] as a consequence of more philosophical constraints:

Definition 4.10. *A maximally even set with cardinality d in* $\mathbb{Z}_n$ *is the set of values of one of the following J-functions:*

$$J_{d,n}^\alpha(k) = \lfloor \alpha + \frac{kn}{d} \rfloor \quad \mod n, k = 0, 1 \ldots d - 1.$$

One can choose the round function instead of the floor function (or ceiling) with equivalent results. This formula approximates exact divisions of n into d parts, which is of course impossible to do exactly unless $d \mid n$.

Example 4.11. Depending on the offset α, the $J_{7,\alpha}^{12}$ generates the 12 major scales (in fifth order), for instance

$$J_{7,12}^0(\llbracket 0, 6 \rrbracket) = \{0, 1, 3, 5, 6, 8, 10\}, \text{ i.e. D}\flat \text{ major,}$$

whereas C major is generated by $J_{7,12}^5(\llbracket 0, 6 \rrbracket)$.

4.2.1 Some regularity features

It is possible to define the class $\mathrm{ME}_{n,d}$ as the generic ME set with d elements in $\mathbb{Z}_n$, because this class is invariant under the action of T/I: any ME set in the class is translated (and also inversed) from any other one.[9] It follows that the number of different ME sets with given (n, d) is a divisor of n, depending on inner periodicities in the set. We will see also that the complement set of a ME set is still a ME set.

An aesthetically remarkable feature of ME sets is the precise quantity of variants of intervals between consecutive elements, or more generally of typed subsets. This is better explained with an example: consider $\{0, 2, 4, 7, 9\} = \mathrm{ME}_{12,5}$. Consecutive intervals, or steps, come in exactly two sizes (2 or 3). The same is true for 'thirds', leaving every odd note out: they are $4 - 0 = 4, 7 - 2 = 5, 9 - 4 = 5, 0 - 7 = 5, 2 - 9 = 5.$[10] Similarly, consecutive triplets like $(0, 2, 4), (2, 4, 7), (4, 7, 9)$ come in three configurations, as do the 'triads' $(0, 4, 9), (2, 7, 0), (7, 0, 4)$ and so on. When this cardinality of a subset of the scale is always equal to the variety of different instances of the type of subset ('Cardinality=Variety'), the scale is said to be Well-Formed, henceforth WF for short. See [28] for much more on this subject. ME sets are WF, or degenerate-WF; for instance the whole-tone scale $\mathrm{ME}_{12,6}$ has only one step size, not two.

One definite advantage of the definition of ME sets in terms of DFT below is that it makes obvious that the complement of a ME set is a ME set. Indeed, from the

---

[9] This will be proved easily with the alternative DFT definition provided below.
[10] Tymoczko points out these 'thirds' in pentatonic context in the last phrase of Debussy's *La Fille aux cheveux de lin*.

typology below or the *J*-function definition one easily gets the following paradoxical statement:

Theorem 4.12. *Let $A \subset \mathbb{Z}_n$ be a ME set and $B = \mathbb{Z}_n \setminus A$ its complement. Then B is a ME set; moreover, some translate of B is included in A or the reverse.*

As I mentioned and proved in [9], this 'Chopin's theorem' holds *mutatis mutandis* for generated scales: when a scale and its complement are both generated, they share their set of generators. This is of course reminiscent of Babbitt's theorem. The reference to Chopin of course alludes to his *Etude op. 10, n° 5*, cf. Fig. 4.7, wherein the pentatonic played throughout the piece by the right hand is a subset of the major scales (mostly G♭ and B♭) played by the left hand.

Fig. 4.7. Pentatonic vs. diatonic

A nice application to rhythms is Astor Piazzolla's use of the complement of tresilo $T = \{0,3,6\} \subset \mathbb{Z}_8$: he uses the pattern $C = \{1,2,4,5,7\}$, not only in its function of complement of T, which is fairly common in post 1950s-tango, but also as a basis for a secondary theme in *La Milonga del Angel*. As discussed in Theorem 4.12, the ternary pulsation is present also in this complement rhythm, see Fig. 4.8 which shows how 'the silence in tango is still tango'.

4.2.2 Three types of ME sets

A fine distinction

In [72], Quinn introduces a typology of ME sets, depending on $m = \gcd(d,n)$. We reproduce this classification here, since it is closely related to questions of inner periodicities and complementarity, qualities that can actually be diagnosed at a glance on the DFT. The seminal case is

Definition 4.13. *A type I ME set happens when $m = 1$. The scale is generated (and WF).*

It is generated by the multiplicative inverse f of d in $\mathbb{Z}_n$, or by $-f$ (these are the only two generators). Typical examples are the diatonic and pentatonic scales in $\mathbb{Z}_{12}$. All *its Fourier coefficients are non zero* (a trivial consequence of Theorem 4.7).

Fig. 4.8. Tresilo and its complement in Piazzolla's *Milonga del Angel*

Definition 4.14. *A type II_a ME set happens when $m = d$, i.e. $d \mid n$. The scale is generated, but it is degenerate WF, dividing $\mathbb{Z}_n$ into a regular polygon.*

Typical examples are the diminished seventh $D7 = \{0, 3, 6, 9\}$ (Fig. 8.10) and whole-tone scale $WT = \{0, 2, 4, 6, 8, 10\}$ (Fig. 8.23). The DFT is quite characteristic: coefficients are 0 except those whose index is a multiple of n/d, which are all equal to d. For instance, for a diminished seventh it is $(4, 0, 0, 0, 4, 0, 0, 0, 4, 0, 0, 0)$.

Definition 4.15. *A type II_b ME set happens when $1 < m = n - d < d$. It is the complement of a type II_a ME set.*

Since the complement has cardinality m, which is a divisor of n, it is ME because the complement of a ME set is a ME set (proved below). The prototype is the octatonic collection $\{0, 1, 3, 4, 6, 7, 9, 10\}$ (Fig. 8.31). Its DFT is the same as type II_a (except of course the 0th coefficient).

Definition 4.16. *Type III ME sets gather the remaining cases: $1 < m < d, m \neq n - d$.*

The DFT is a compound of the two other types: the varied values of the DFT are the same as in type I, with 0's interspersed because of its periodicity (remember the formula for oversampling, cf. Fig. 1.5). For instance $\{0, 2, 4, 6, 9, 11, 13, 15\} = ME_{18,8}$ (Fig. 4.9) yields coefficients (magnitudes)

$$(8, 0, 1.06, 0, 1.3, 0, 2, 0, 5.76, 0, 5.76, 0, 2, 0, 1.3, 0, 1.06, 0).$$

This classification in three types is stable by complementation.

The last two classes are ME sets with a smaller period, i.e. what Messiaen called Limited Transposition Modes. They are all concatenated from smaller ME sets.

Remark 4.17. Clampitt *et alii* [28] argue that type I is fundamental, inasmuch as this type generates all others: type III is obtained by slicing n into m equal parts and filling each part with the same type I ME set with d' notes among $n' = n/m$, see Fig. 4.9.

Remark 4.18. Types II and III are 'perfectly balanced' in the sense of [67], i.e. $a_1 = 0$ (they are unions of regular polygons). Note that this perfect balancing, a pure Fourier quality, fails to characterise ME sets: for $ME_{(12,7)}$,

$$|a_1| = \frac{\sin(\pi/12)}{\sin(7\pi/12)} = 2 - \sqrt{3} \approx 0.26795$$

is not the smallest value for seven-note scales, superseded by $\{0,1,2,5,6,8,9\}$ for which $a_1 = 0$, cf. Fig. 8.28.[11]

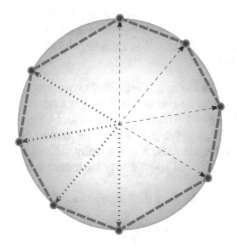

Fig. 4.9. A type III ME set : $\{0,2,4,6\} \oplus \{0,9\} \subset \mathbb{Z}_{18}$

Existence of type III ME sets

Quinn ([72]) was remarkably astute in this taxonomy, since as he himself pointed out there are no type III ME sets when $n = 12$, a rather prominent case for West-European music at least. This type exists though: for instance, when $n = 18$, consider $ME_{(18,8)} = \{0,2,4,6,9,11,13,15\} = \{0,2,4,6\} \oplus \{0,9\} = ME_{(9,4)}$ redoubled, shown in Fig. 4.9. Incidentally, its DFT can be computed easily from this decomposition, since the DFT of $\{0,2,4,6\}$ in $\mathbb{Z}_9$ is (in magnitudes)

---

[11] The fifth coefficient is also nil, since this balanced scale type is invariant by affine transformations: $5 \times \{0,1,2,5,6,8,9\} = \{0,1,2,5,6,8,9\} + 4 \mod 12$.

$$(4, 0.53, 0.65, 1., 2.88, 2.88, 1, 0.65, 0.53)$$

and it only remains to intersperse zeroes and multiply by 2 to retrieve the DFT of $ME_{(18,8)}$ already given above.

Of course type III is impossible when n is prime, since in this case only type I happens (barring the full aggregate or the empty set). But for large composite n, type III is always possible:

Theorem 4.19 (Amiot, 2005).
For composite $n > 12$, there exists d such that $ME_{(n,d)}$ has type III.

The proof hinges on a technical

Lemma 4.20. *For composite $n > 12$, there exists $k \mid n$ and a prime number $p < k-1$ such that p is not a divisor of k.*

Proof. Notice that for $n = 12$ the lemma fails, since at most $k = 6$ and all prime numbers $p < 5$ divide 6.

Consider a composite $n \geqslant 25$ – lower values are checked by hand or computer. The general idea is to have k be the largest strict divisor of n. It can be written either $k = 2m+1$ or $k = 2m+2$. Since n/k is a smaller divisor of n, $k \geqslant n/k$, i.e. $k \geqslant \sqrt{n}$, hence $k \geqslant 5$ and $m \geqslant 2$.

- First case: $n = 2^r$. Let $k = n/2, p = 3$. Works whenever $n \geqslant 8$.
- Second case: n admits an odd divisor $k \geqslant 5$, not necessarily prime. Select this value for k, and let $p = 2$. This works for $n = 10, 14, 15 \cdots$.
- Last case: $n = 2^a 3^b, a \geqslant 1, b \geqslant 1$. This is the trickier case, since it is for $n = 2 \times 2 \times 3$ that the lemma fails. It is not really difficult though, since whenever $n \geqslant 24$, setting $k = n/2$ and $p = 5$ satisfy the lemma conditions.

The theorem follows now from the construction

$$j \mapsto \lfloor \frac{kj}{p} \rfloor = \lfloor \frac{jn}{np/k} \rfloor$$

yielding a type III ME set, concatenated from $ME_{k,p}$ which is a type I in $\mathbb{Z}_k$ since p does not divide k.

4.2.3 DFT definition of ME sets

This definition is our principal aim in this section: Quinn discovered that ME sets can be characterised by a high value of some Fourier coefficient. To quote [72]:

> We note that generic prototypicality may be interpreted as maximal imbalance on the associated Fourier balance – at least to the extent that a generic prototype tips its associated Fourier balance more than any other chord of the same cardinality possibly can.[12]

---

[12] Quinn was originally interested in what he calls 'prototypical chords', defined by cultural consensus, and which happen to be ME sets.

More precisely, as proved rigorously in [10] with excruciating detail, one can adopt the following definition as equivalent to the other ones (say Def. 4.10):

Definition 4.21. *The pc-set $A \subset \mathbb{Z}_n$, with cardinality d, is a ME set if the number $|\mathscr{F}_A(d)|$ is maximal among the values $|\mathscr{F}_X(d)|$ for all pc-sets X with cardinality d:*

$$\forall X \subset \mathbb{Z}_n, \quad \#X = d \quad \Rightarrow \quad |\mathscr{F}_A(d)| \geqslant |\mathscr{F}_X(d)|.$$

From the formulas already derived for DFT, it follows without further ado

Proposition 4.22. *Transposition, inversion and complementation of a ME set still yield a ME set: any pc-set homometric to a ME set is a ME set.*

This is obvious since all these operations preserve the magnitude of Fourier coefficients, which is a definite advantage over alternative definitions. It also hints that the magnitude of Fourier coefficients might be a perceptible quality – at least it is one commonly recognised.

We will show that the DFT definition is equivalent to the definition pinpointing a generated scale, in the spirit of Rem. 4.17. Reduction to the *J*-function definition has been carried in [10] and would be redundant here, since the equivalence of all previously known definitions had been already proved in seminal works on ME sets.

Proof. Quinn provided a simple argument which is fairly convincing for the type I case when $\gcd(d,n) = 1$, and even more in the degenerate case – but insufficient for the remaining cases. Remember

$$\mathscr{F}_A(d) = \sum_{k \in A} e^{-2idk\pi/n} = \mathscr{F}_{dA}(1)$$

where dA may be a multiset.

When $d \mid n$, one easily gets $\mathscr{F}_A(d) = d$ for $A = \{0, n/d, 2n/d \dots\}$, a regular subdivision of $\mathbb{Z}_n$. Conversely, one has $|\mathscr{F}_A(d)| \leqslant 1 + 1 + \dots 1 = d$ by triangular inequality, and the equality (for a Euclidean norm) may only happen when the complex exponentials involved all point to the same direction, since $|z + z'| < |z| + |z'|$ for non-colinear z, z'. But this happens if and only if

$$\forall k, k' \in A \quad 2dk\pi/n = 2dk'\pi/n \quad \mod 2\pi \iff k = k' \quad \mod (n/d);$$

hence (since $\#A = d$) A is a whole arithmetic sequence with common difference n/d.[13] In this case, $A' = dA$ is a multiset with exactly one element repeated d times.

When $\gcd(d,n) = 1$, multiplication by d is bijective and $A' = dA$ is a genuine set with the same cardinality as A. All the exponentials must then be distinct, so the argumentation above does not work. Quinn argues that these exponentials should be as close as possible one to another[14], meaning that A' is a chromatic cluster $\{1, 2 \dots d\}$. This can (and should) be formally proved using

---

[13] The same argument proves that for $\#A < d$, $|\mathscr{F}_A(d)|$ will be maximal when A is a subset of such a sequence, see Section 4.3.

[14] 'The best the chord can do is to have pcs gathered in adjacent pans, so that the arrows point in approximately the same direction', ibid.

Lemma 4.23 (Huddling together).

Have d points $a_1, \ldots a_d$ on the unit circle S^1, and move a_1 towards the sum $s = \sum_{k=1}^{d} a_k$, meaning a_1 is replaced by a_1' whose argument (or phase) is between the phases of a_1 and s.

Then $|a_1' + a_2 + \ldots a_d| \geqslant |a_1 + a_2 + \ldots a_d|$.

Proof. In a nutshell, the sum increases because the angle between $s = \sum_{k=1}^{d} a_k$ and $a_1' - a_1$ is acute. Let us provide a comprehensive computation: up to rotation and symmetry, one can assume without loss of generality that $\arg(s) = 0$ and $\varphi_1 = \arg(a_1) \in\,]0, \pi]$; then $\varphi_1' = \arg(a_1') \in [0, \arg(a_1)] \subset [0, \pi]$ so a_1 and a_1' are both 'above', see Fig. 4.10.

Since cos is decreasing on $[0, \pi]$ we have

$$\cos \varphi_1 + \cos \varphi_2 + \ldots \cos \varphi_n \geqslant \cos \varphi_1' + \cos \varphi_2 + \ldots \cos \varphi_n.$$

These sums are the projections of s and $s' = a_1' + a_2 + \ldots a_d$ on the real axis. But s is assumed to be real, and $|s'|$ is greater than its projection. Hence $|s'| \geqslant s$ and more precisely $|s'| > s$ unless $\varphi_1 = \varphi_1'$.

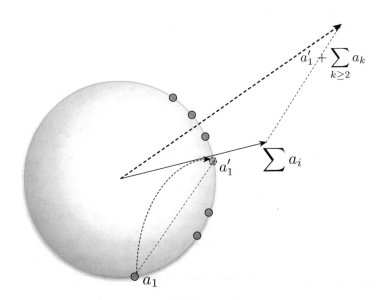

Fig. 4.10. The length of the sum increases.

The fact that A' must be a chromatic cluster follows: else, A' would feature holes in the sequence between its elements[15], and one extremal point could be moved to

---

[15] Writing A' in a 'basic form' such as $A' = \{0, \alpha, \beta \ldots \omega\}$ with $0 < \alpha < \beta < \cdots < \omega < n$ and $n - \omega$ maximal, for instance.

one such hole, increasing $|\mathscr{F}_{A'}(1)|$ in the process. This can be iterated until we get a chromatic cluster and no more.

Since $dA = A' = \{1, 2, \ldots d\}$ or some translate thereof, we find $A = fA' = \{f, 2f, \ldots df\}$ where f is the multiplicative inverse of d in $\mathbb{Z}_n$. In the seminal example, the diatonic collection with 7 elements is generated by fifths since $7^{-1} = 7$ mod 12. The previous discussion on the number of generators of a generated sequence modulo n shows that in this case there are only the two generators f and $-f$.

The remaining case $\gcd(d, n) > 1$, with d not a divisor of n, is slightly more complicated. Let $m = \gcd(d, n), n' = n/m, d' = d/m$: then n' and d' are coprime and we aim at reducing the study to the preceding case. *For instance, consider the case of $A = \{0, 1, 3, 4, 6, 7, 9, 10\}$ (the octatonic collection) with $d = 8, m = 4, n' = 3, d' = 2$.* Indeed $\pi_d : x \mapsto dx$ now maps $\mathbb{Z}_n$ to $\mathbb{Z}_{n'}$, each fiber (pre-image) having m elements. Assume $|\mathscr{F}_A(d)|$ is maximal and let $A' = \pi_d(A)$ (here we consider A as a set, not a multiset. See [10] for a proof in the context of multisets). Then

$$\mathscr{F}_A(d) = \sum_{k \in A} e^{-2ikd\pi/n} = \sum_{k' \in A'} m(k')e^{-2ik'd'\pi/n'} = \sum_{k'' \in A'' = d'A'} m(k')e^{-2ik''\pi/n'}$$

where $m(k') = \#(\pi_d^{-1}(\{k'\}))$ denotes the cardinal of the fiber, i.e. the number of times k' is hit as an image of an element of A. Lemma 4.23 can be used here since it does not assume the points to be distinct. We can huddle the elements of $A'' = d'A'$ up to m times each, since $m(k') \leqslant m$. Hence in the maximal case, A' has d' elements, each fiber contains m antecedents, i.e. A is periodic since for any $a \in A$ we must have all the l different $a + k\frac{n}{m} \in A$ *(for the octatonic example, A'' is $\{0, 4\} \subset 4\mathbb{Z}_{12} = \{0, 4, 8\}$ with each element repeated four times)*; hence

$$|\mathscr{F}_A(d)| \leqslant m|\mathscr{F}_{A'}(d')| \leqslant m \max_{B \subset \mathbb{Z}_{n'}, \#B = d'} |\mathscr{F}_B(d')|.$$

For the maximal value to be reached, A' must be maximally even (i.e. the elements of A'' form a chromatic cluster) and each fiber must be full (i.e. each $m(k')$ is equal to m, meaning A is the whole of $\pi_d^{-1}(A')$). This means

Proposition 4.24. *In the case $m = \gcd(d, n) > 1$, d not a divisor of n, a set $A \subset \mathbb{Z}_n, \#A = d$ is maximally even iff $A' = dA$ is maximally even in $\mathbb{Z}_{n/m}$ and A is m-periodic. In other words, A must be concatenated from A'.*

In the example proposed, $A' = \mathrm{ME}_{3,2}$ – for instance $A' = \{0, 1\} \in \mathbb{Z}_3$ – and hence $A = \pi_4^{-1}(A') = \{0, 1, 3, 4, 6, 7, 9, 10\} = A' \oplus 3\mathbb{Z}_{12}$ with a slight abuse of notation.

This description of the last case exemplifies the transfer of the DFT from A to its projection on an appropriate subgroup of $\mathbb{Z}_n$, cf. Proposition 3.36 above. It is illuminating to compare the DFTs of A and A' in Fig. 4.11, where a simple scale change allows us to superimpose both graphics.

To sum it up, the Fourier definition of ME sets pinpoints the quality of being as close as possible to a regular subdivision of the circle – etymologically, a cyclotomy.

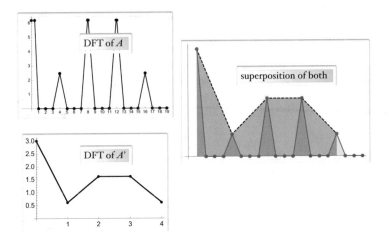

Fig. 4.11. Fourier magnitudes of a periodic ME set and its type I projection

4.3 Pc-sets with large Fourier coefficients

4.3.1 Maximal values

We have just seen that $|\mathscr{F}_A(d)|$ is maximal for $\mathrm{ME}_{n,d}$, among all d-subsets. One may well ask what are the maximal cases for other coefficients. For instance, when one keeps the cardinality d fixed, the pc-sets which maximise $|a_1|$ are the chromatic clusters, e.g. $\{0,1,2\ldots d-1\}$ as we have established during the proof of the type I-ME set case.

An extension of this result yields the maximum case for $|\mathscr{F}_A(k)|$ when k is coprime with n: in this case kA is a set, not a multiset and

$$\mathscr{F}_A(k) = \mathscr{F}_{kA}(1)$$

is maximal when kA is a chromatic cluster, meaning that A is generated, with generating interval k^{-1}, the inverse of k in $\mathbb{Z}_n$. As a corollary, all maximum values of $\mathscr{F}_A(k)$ are identical for $k \in \mathbb{Z}_n^*$. For fixed d, this maximum is $\frac{\sin(d\pi/n)}{\sin(\pi/n)}$ (which gets close to d when n is large).

Remember that k, k' are *associated* if there exists $\lambda \in \mathbb{Z}_n^*$ such that $k' = \lambda k$. Then we can generalise slightly the above computation:

Proposition 4.25. *The maximum of $|\mathscr{F}_A(k)|$ on d-subsets is the same as the maximum of $|\mathscr{F}_A(k')|$ for k' associated to k.*

However, it is a completely different case when $\gcd(k,n) > 1$, because kA can then be a multiset, not a set, as we have seen for type II and III ME sets. It may even be possible to reach $|a_k| = d$, for type II ME sets or their subsets. This happens whenever $d \leqslant \frac{n}{\gcd(k,n)}$.

Example 4.26. Any subset of a whole-tone scale has maximum $\mathscr{F}_A(6)$: for instance for $A =$ CDF♯G♯ $= \{0,2,6,8\}$, $\mathscr{F}_A(6) = 4 = \#A$, cf. Fig. 8.11.

The most complicated cases are reminiscent of the study of saturation in one interval: sometimes d is larger than all strict divisors of n. Of course, if $d > n/2$ we already know that the Fourier coefficients are the same as those of the complement subset, so let us assume $d < n/2$ (the case $d = n/2$ yields a maximum $\mathscr{F}_A(d) = d$ for $A = 2\mathbb{Z}_n$). Following the general idea of the proof of the DFT definition of ME sets, we want the multiset kA to be as huddled as possible: if repetition of a single value is not available, then we aim for repeating several huddled values. This happens when kA is a repetition of a subset of a regular polygon, with the eventual added points all situated on the same location, see Fig. 4.12.

Example 4.27. Consider $n = 75, d = 27 > n/3$. We can construct a perfect ME set with 25 elements, $A = \{0,3,6\ldots72\}$. Then for $k = 3$ one gets $A'_{mult} = 25A = \{0^{\#25}\}$, i.e. 0 repeated 25 times. Since there is no way[16] to enlarge A without adding new elements to A'_{mult}, the best one can do is to have these extraneous elements in A'_{mult} stay as close as possible to 0. For instance, one can add 4 and 31 to A, which turns A'_{mult} to $\{0^{\#25}, 25^{\#2}\}$, i.e. 0 25 times and 25 twice. The resulting set yields the maximum possible value of $|\mathscr{F}_A(25)|$ for 27-subsets of $\mathbb{Z}_{75}$.

It is not clear that this value is the greatest possible of $|\mathscr{F}_A(k)|$ for 27-subsets and any k. Indeed one has to check for other divisors of 75. In Fig. 4.12, I tried also B, saturated in interval 5, made of a 15-polygon and another, incomplete one as close as possible; and C, saturated in interval 15, union of five pentagons and two points on a sixth; and checked the values of the corresponding Fourier coefficients. In this case, $|\mathscr{F}_A(25)| = 24.062, |\mathscr{F}_B(15)| = 22.506$ and $|\mathscr{F}_C(15)| = 21.206$; hence A achieves the highest possible maximal value of a Fourier coefficient among all 27-subsets of $\mathbb{Z}_{75}$. For the record, $\mathscr{F}_M(27) = 21.658$ for $M =$ ME$_{75,27}$, i.e. the ME set only beats C.

The general question now arises: for a given pair (n,d), what are the subsets $A \subset \mathbb{Z}_n$ with cardinality d that yield the maximal value of their largest $|\mathscr{F}_A(k)|$? There are three cases, summed up by the following:

Theorem 4.28. *Among d-subsets of $\mathbb{Z}_n$ (with $d < n/2$), the sets with the largest Fourier coefficients are*

1. *Subsets of regular polygons (when d is smaller than some divisor of n).*
2. *Maximally even sets.*
3. *The kind of saturated/huddled subsets shown by the example above.*

Notice that even in the last case, some solutions can be generated by J functions. For instance $(0,6,12,1,7,13,2,8)$ in Fig. 4.13 is the sequence of values of $\lfloor 6.34k \rfloor$ mod 18 for $k = 0\ldots7$; indeed even the tango/habanera pattern $\{0,3,4,6\}$ can be achieved as values of $\lfloor 2 + 2.5k \rfloor, k \in [\![1,4]\!]$.[17]

---

[16] If A is a true set, not a multiset.

[17] Keep in mind however that some pc-sets cannot be generated in this way, for instance $\{0,1,4\}$ when $n \geqslant 10$.

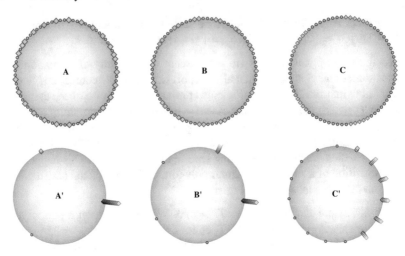

Fig. 4.12. Three *candidates* for maximum max $|\mathscr{F}_A|$ for 27-subsets of $\mathbb{Z}_{75}$

This was first analysed in the third online supplementary of [10]. The last case is somewhat messy: there is no simple formula (one has to check for k being any divisor of n, because the largest divisor does not always yield the highest Fourier coefficient) and the result is not unique up to isometry, in contrast to the ME set cases. The three different cases are exemplified in Fig. 4.13 with $n = 18$ and $d = 7, 5, 8$. The corresponding multisets are shown underneath.

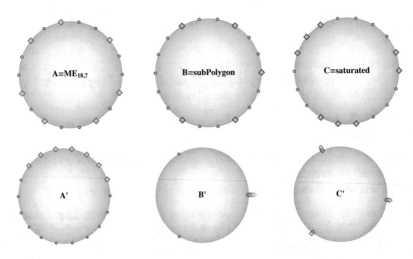

Fig. 4.13. Three *types* for maximum max $|\mathscr{F}_A|$.

When do large values occur?

All these results vindicate Quinn's notion of saliency, i.e. large a_5 show a large *fifthishness* (which we will rename *diatonicity* in the musical examples below) while large a_6 exemplifies *whole-tonedness*, etc. We have already explored the maximal cases, in the end of the discussion let us relax the condition to 'relatively large' with, of necessity, fuzzier assertions.

Example 4.29. For instance, for a short excursion in the rhythmic domain we can assert that the tresilo $(0,3,6)$ in $\mathbb{Z}_8$ has maximal 'ternariness', i.e. largest a_3 among all 3-sets ($|a_3| = 2.41$). But the standard tango pattern $(0,3,4,6)$ in the same $\mathbb{Z}_8$ has some ternary saliency too ($|a_3| = 1.85$), though its largest Fourier coefficient is the fourth ($|a_4| = 2$), asserting that tango music is binary though with a strongish ternary intent.[18] The four-note rhythm with best ternary saliency is $(0,1,3,6)$, a generated set generalizing the ME-sets construction:

$$\{0,1,3,6\} = \{0,3,6,9\} = 3 \times \{0,1,2,3\} \quad \mathrm{mod}\ 8 \quad (|a_3| = 2.61).$$

It has been observed [25, 98] that frequent occurrences of some intervals between pc-sets (measured on a time span of one to five bars of the score, for instance) are correlated with large values of some Fourier coefficients – the fifth interval with the fifth coefficient, or minor thirds with the fourth coefficient, for instance. This is well in line with what we discussed in Section 4.1.1, and easier to adapt than the notion of maximal evenness. Is it a really reliable guideline though?

Example 4.30. Since it is a periodic ME set, $\mathcal{O} = \{0,1,3,4,6,7,9,10\}$ (the octatonic collection) has clear-cut Fourier coefficient magnitudes: $|\mathscr{F}_{\mathcal{O}}| = (8,0,0,0,4,0,0,\dots)$. The zeroes reflect the periodicity of this pc-set (the coefficients from 7th to 11th have been omitted since their values are reversed from the first ones).

Subsets of this collection still preserve the *saliency* of the fourth coefficient: for $A = \{0,1,3,4,7,9\}$, one finds $(6,1,1,2,\mathbf{3},1,2,\dots)$ and for $A' = \{0,1,3,4,6,7\}$, $|\mathscr{F}_{A'}| = (6,1.93,1.73,1.41,\mathbf{3},0.52,0,\dots)$.[19]

The last two examples both display four minor thirds, and though the fourth Fourier coefficient has the same magnitude, the other coefficients do not. The more we stir away from the regular subsets studied before, the less exact the correlation between saturation and saliency becomes, cf. 5.4 below.

For generated sequences whose generator is not a divisor of n, or bouts of such sequences which are not ME sets, first remember that a generated sequence features more occurrences of the generating interval than several juxtaposed partial sequences: there are six second intervals in a whole-tone scale $WT = \{0,2,4,6,8,10\}$, but only four in the Guidonian hexachord $GH = \{0,2,4,\mathbf{5},\mathbf{7},\mathbf{9}\}$ which is a reunion

---

[18] Indeed a kind of walz, *El vals criollo* is among the three principal styles of music played and danced in tango balls.

[19] This somewhat informal remark is very important, as it will lead us to replace advantageously the 'complex' manipulations in Forte's 'Set Theory' (i.e. subset relationships) by consideration of saliency. This is a *forte* of DFT theory, noticed by Yust.

of two three-note whole-tone sequences (i.e. a convolution product of $\{0,2,4\}$ by $\{0,5\}$, see Figs. 8.8 and 8.5 respectively). On the other hand, this last pc-set is a full-fledged fifth sequence $(5,0,7,2,9,4)$. All this appears clearly on the Fourier magnitudes, see also Fig. 8.26 and 8.23:

$$|\mathscr{F}_{WT}| = (6,0,0,0,0,0,6,\ldots) \qquad |\mathscr{F}_{GH}| = (6,1.035,0,1.414,0,3.864,0,\ldots).$$

Notice that the sixth coefficient, maximal for WT, altogether vanishes in GH despite the four whole tones in it[20] which shows crudely that the magnitude of a Fourier coefficient is not completely equivalent to the frequency of occurrence of a corresponding interval. However, in tonal music where a diatonic universe is often prevalent, the organisation of fifths often adheres to the generating sequence of the diatonic, which is maximal in number of fifths, and the 5th Fourier coefficient is accordingly large – as we have seen in Section 4.2, the diatonic collection has maximum magnitude $(1+\sqrt{3} \approx 2.73)$ among all other seven-notes pc-sets for the fifth coefficient. Its most frequent subsets, the simple and popular boogie/rock bass sequence CFG (057) and the pentatonic collection, reach exactly the same value. In the former case (CFG) this is not far from the absolute maximum possible for the DFT of a 3-pc-set. In the latter we have the absolute maximum.

So when can we rely on the informal remark above, since it is not always true?

The Fourier transform being continuous, slight modifications of a pc-set entail slight modifications of the Fourier coefficients. Hence the somewhat vague, but informative, assertion:

Proposition 4.31. *Usually, pc-(multi)sets with a high frequency of occurrence of interval d are close to (subsets of) arithmetic sequences with generator d and yield a high value of their k^{th} Fourier coefficient, where k is*

- *n/d when d divides n, or*
- *$d^{-1} \in \mathbb{Z}_n$ when n, d are coprime.[21]*

This lacks a precise definition of 'closeness' to a given pc-set, a notion that is open to interpretation, and leaves aside the case of a loose relationship between d and n (neither divisor nor coprime). It is also debatable for small d and especially $d = 1$, though there is some correlation in this case with the number of *successive* semitones but their overall distribution could ruin this character, see Fig. 8.28 where a scale with four semitones has $a_1 = 0$.[22]

We will discuss in Section 5.4 a relationship between size of DFT coefficient and voice-leading distance to a (usually virtual) chord with maximum value, first estimated by Tymoczko and improved for the present publication.

---

[20] This is because there are as many odd pcs as even. Another way to look at it is that this coefficient is nil already for the *factor* $\{0,5\}$.

[21] See Section 4.2 for an explanation of this value of k.

[22] The only such seven-note scale.

4.3.2 Musical meaning

A word of caution is in order: when considering the character of a pc-set (diatonic, whole-tonic, etc...) we usually compare the respective magnitudes of appropriate Fourier coefficients. But it could well be argued that these magnitudes should be weighted: *for instance, coefficient a_2 can be as large as 6 (for a whole-tone scale) but a_5 (or a_1) is never more than $\sqrt{2}+\sqrt{6}$ (Guidonian hexachord). However, these limitations fall when one drops genuine pc-sets and considers continuous DFT, even if the musical notions underlying, say, a regular division in seven of an octave, are more virtual than real. In balancing these arguments, I prudently chose not to choose and left the comparison of magnitudes of Fourier coefficients as is, though perhaps with a modicum of salt. For instance, the jingle for* la Société Nationale des Chemins de Fer *created by Michael Boumendil (which I quote because David Gilmour, Pink Floyd's lead guitarist, fell in love with it and used it as a leitmotif in his song* Rattle That Lock*: see* https://www.youtube.com/watch?v=L1v7hXEQhsQ*) arpeggiates a seventh chord CGA♭E♭; the corresponding profile in Fig. 8.12 shows a large a_3, i.e. 'major thirdishness' or 'augmentedness', which indeed correlates with the presence of three thirds (two major, one minor). But the value of $|a_5|$, though only 2/3 of $|a_3|$, is comparatively large because it is closer to the maximum theoretical value for a_5 (indeed, the pc-set is almost saturated in fifths), and hence the pc-set is also fairly diatonic, which is good for rock music.*

The six characters

We may as well begin with clarifying the meaning of saliency for coefficients 1,2,3,4,5,6 in $\mathbb{Z}_{12}$. I take them from the easiest to the less obvious. Examples are provided on Fig. 4.14.

Fig. 4.14. Examples of the six characters

- The sixth is easiest to understand, especially using Quinn's (weighing) 'scales': this coefficient is greater when its pcs concentrate in one of the two whole-tone scales. It is uncontrovertibly the *whole-toneness*. Clear-cut examples can be found on Figs. 8.23, 8.15, 8.23; a more ambivalent case is the Guidonian hexachord in Fig. 8.23, a reunion of two whole-tone tetrachords CDE - FGA, but with opposite polarities resulting in zero 'whole-toneness'.

- As we have already discussed at length, the fifth coefficient can well and truly be called the *diatonicity* of a pc-set: it has everything to do with the tonal character (or alternatively the generatedness by fifths) which marks pentatonic and diatonic scales among other prominent specimens, see Fig. 8.18 and 8.29 or even 8.7 (CFG) and 8.8 (CDE).[23] Notice that the rather large index 5 discriminates dramatically between just and diminished fifths, since a tritone has nil a_5 but a fifth is the maximal dyad for this saliency, cf. Example 4.32 below.
- Third and fourth mark on the one hand generatedness (or saturation) in major and minor thirds respectively, but *minor(major)-thirdishness* is a somewhat ambiguous notion: among subsets with similar cardinality, any subset of a diminished seventh features a maximal magnitude for a_4, but so does the octatonic scale (among eight-notes pc-sets); and I can agree with J. Yust who dubs *octatonic* the pc-sets with large a_4 – they are usually subsets of some octatonic scale. As for *major* thirdishness, I like to think of it as 'augmentedness', good prototypes being the augmented triad or the 'magic' hexachord $\{0,1,4,5,8,9\}$ (also called 'ode to Napoleon'), cf. Fig. 8.24.
- From the discussion above, one could wonder whether a large a_1 corresponds to many semitones or many (major) sevenths, but the issue is not large and we will call *chromatic* any pc-set with a comparatively large a_1. However it should be noticed that too many notes will perforce diminish this coefficient. For instance, the scale B C D♭ E F G A♭ or $\{0,1,4,5,7,8,11\}$ has $a_1 = 0$ (see [67] and Fig. 8.28) though it features many semitones. Notwithstanding, decent prototypes are chromatic chunks of lengths 4 to 6, i.e. Figs. 8.14, 8.22, 8.27 with chromaticities equal respectively to 3.35, 3.73, 3.86. These coefficients are less sensitive than a_5: a major triad is generated neither by major nor minor thirds but both coefficients a_3, a_4 (and of course a_5 too) are fairly large.
- The more troublesome coefficient is a_2. Yust uses Messiaen's Limited Transposition Mode $M_5 = \{0,1,2,6,7,8\}$ as a prototype (Fig. 8.25), together with $M_4 = \{0,1,2,3,6,7,8,9\}$ (Fig. 8.30) which sports almost the same value (and is more frequently used, if only in R. Wagner's *Tristan*, cf. [5]). I like the neologism *tritonic* to qualify pc-sets with large a_2, though Yust's *quartal quality* is convincingly expostulated in his example of Ruth Crawford Seeger's 'White Moon' [98, 100].[24] It is perhaps an artifact of working modulo 12, but as he points out, this quality quite often goes with a lack of thirds and sixths, which is a hallmark of some early 20th century music: for instance, the prominence of this coefficient in B. Bartok's Fourth Quartet can be arguably correlated to its acknowledged 'modernism' [98].[25]

This classification makes it really easy to appreciate the character of any given pc-set:

---

[23] Actually the Guidonian hexachord does slightly better than all other pc-sets, with 3.86 instead of 3.73 for the diatonic, a minor triumph for archeo-musicologists perhaps.

[24] Sandburg Songs, n° 2.

[25] And of course, if one Fourier coefficient is large, then the others are left less room, since the sum of their squares is fixed.

Example 4.32. Consider for instance the three aggressive fifths initiating C. Debussy's *La Puerta del Vino* (Préludes, II): A♭-D♭, E-A, A-D, constituting the pc-set $\{1,2,4,8,9\}$. Its Fourier profile can be found online or computed with the software I provided with this book, or even roughly estimated: to begin with, the tritone $(2,8)$ can be cancelled out for any odd-indexed coefficient, leaving $\{1,4,9\}$ to be exponentiated and summed with diverse coefficients. Since this is very close to an equilateral triangle ($\{1,5,9\}$) the coefficient a_1 must be quite small, i.e. the pc-set is not chromatic (character 1). On the other hand, multiplying by 5 yields $\{5,8,9\}$, whereas a maximum would be reached for $\{7,8,9\}$; hence our pc-set is somewhat diatonic (bearing in mind though that only three of its five notes bear their weight on this character). Most of the bulk is carried by the quartal quality: multiplying by 2 yields the multiset $\{2,4,4,6,8\}$ (after reordering) whose vector sum has magnitude close to 3 after cancellation of the tritone $(2,8)$. The remainder is on the 'augmented' quality, i.e. a_3, which can be computed from multiset $\{3,0,3\}$.

All in all, this describes a non-chromatic, still diatonic but fairly modern pc-set, which I would say is an accurate description of a listener's intuitive perception. Check its profile in Fig. 8.19.

Examples in modal music

Fig. 4.15. *Voiles*, Preludes vol I, C. Debussy

In *Voiles*, C. Debussy opposes quite stringently two of those pure archetypes: the whole-tone scale, which is used for most of the piece, and the pentatonic (black keys) which occurs during the climax just before the last page, back in whole-tone (Fig. 4.15). A deaf scientist, riveted to the meters of Fourier coefficients during the piece, could not miss the exchange of high values between a_6 (from concentration in 6 down to 1) and a_5 (from $2 + \sqrt{3} \approx 3.7$ down to 0) coefficients, even without any knowledge of scales and music theory, cf. Fig. 4.16. The Fourier profiles are provided separately in the tables, Figs. 8.23 and 8.18.

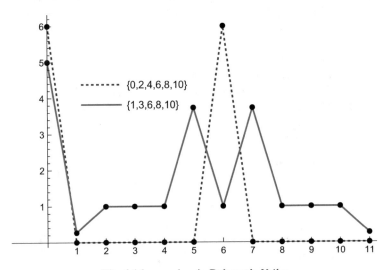

Fig. 4.16. a_5 and a_6 in Debussy's *Voiles*

This is caricatural of course, since traditional analysis of scale content gives the same result quite easily. The point of Fourier analysis of saliency is that it can help decide the character of (a passage of) a piece in less clear-cut cases. Less caricatural perhaps is the pivotal oscillation between the first and middle section, playing on the intersection of the two pc-sets $\{6, 8, 10\}$, i.e. G♭A♭B♭: again set-theoretic considera-tion provides an adequate explanation of this move (around A♭ which happens to be a common center of symmetry of both scales), but it does not hurt to recall that the pc-set $\{6, 8, 10\}$ is both whole-tonic and diatonic, i.e. with both a_5, a_6 large, see the central peak in Fig. 8.8.

A more striking example of the efficiency of DFT magnitude is Yust's analysis in [98] of the beginning of Bartok's Fourth String Quarter IV's movement iv, wherein the melody plays an acoustic scale opposing the accompaniment on DE♭GA♭, see fig. 4.17.

A much more detailed analysis is to be found in the reference given. Here we will simply observe that classical comparison of these two pc-sets is difficult, and

Fig. 4.17. Bartok's String Quartet 4, iv, mm 6-12

that the analyst is tempted to resort to subjective qualities of the scales involved, while confrontation of the DFT's magnitudes is illuminating, see Fig. 4.18.

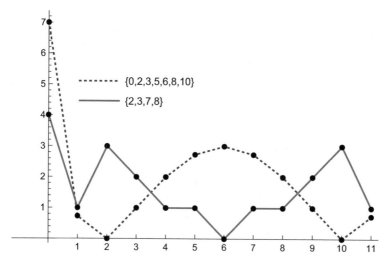

Fig. 4.18. DFT magnitudes of melody and accompaniment in Bartok's String Quartet 4, iv, mm 6-12

The second component, the quartal or tritonic quality, is nil for the acoustic scale in the melody, but large for the bass. Indeed, the latter shows a decidedly atonal quality. Conversely, the whole-toneness of the acoustic scale is large ($|a_6| = 3$) while the accompaniment's is nil, its four notes being equally distributed in the two whole-tone scales, i.e. Quinn's two 'pans'. The values of the first (chromaticity) and fifth (diatonicity), while not as contrasted as second and sixth, are also very revealing of the opposite characters of melody and accompaniment.

Magnitude of Fourier coefficients can help resolve old conflicts. In [100], Yust observes three clearly diatonic voices in Stravinsky's *Three Pieces for String Quartet*, first movement, namely GABC, C♯D♯EF♯ and CD♭E♭ (cf. Fig. 4.19 and their Fourier profiles respectively on Figs. 8.17, 8.16 and **??**), whose large coefficents a_5 more or less cancel each other out when the pc-sets are reunited, according to their

Fig. 4.19. Pc-content of three instruments in Stravinsky's *Three Pieces for String Quartet*, first movement

balanced phases around the circle[26], as formulated in Proposition 6.2; the union of these three pc-sets is CC♯D♯EF♯GAB, close to an octatonic collection (see profile in Fig. 8.32 with its spikes on a_4, even more pronounced if one unites the three voices in a multiset, not a set). The octatonic character of this movement is confirmed by the magnitude of a_4 for the second violin and cello.

Traditional set-theoretic analysis (using subset relationships or 'historical' arguments) of this passage and numerous others had so far spectacularly failed to achieve unanimity, see the fur fly between [92], [93] and more recently [91], [84] (I borrow these references from Yust). This kind of issue can now very easily be resolved, simply by measuring $|a_4|$, or looking it up on the Fourier profiles of pc-sets in Stravinsky's scores.

Tropes

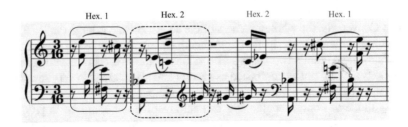

Fig. 4.20. First phrase of Webern op. 27, n° 1 (retranscribed)

---

[26] We will see that the phase of a_5 locates a pc-set on the circle of fifths.

A very common occurrence in dodecaphonic music is the division of the twelve tones in two hexachords, or 'tropes'. It is a golden opportunity to use Babbitt's theorem, either from the intervallic point of view, or using DFT. This is effective on almost any example, such as the first bars of Webern's first movement of Variationen op. 27 (Fig. 4.20).[27] The first hexachord is divided between the two hands[28] in two trichords, FEC♯ and BF♯G. The DFT magnitudes of both trichords and their reunion (dotted line) are shown in Fig. 4.21.

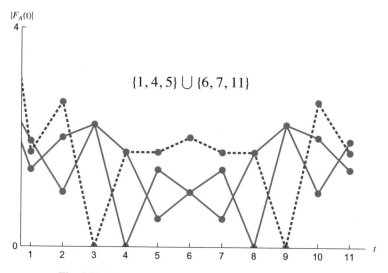

Fig. 4.21. Decomposition of a hexachord in Fourier space

The most interesting coefficient is the third, which is the highest valued for both trichords though it is nil for the hexachord. This brings us a taste of the analysis of directions, or phases, of Fourier coefficients, that will be developed in Chapter 6.

Remember that a_3 is about the 'major-thirdishness' (or 'augmentedness'): specifically, FEC♯ alias $\{1,4,5\}$ has a large $a_3 = 1 - 2i$ (with magnitude $\sqrt{5}$) that points towards the closest pc-*multi*set with maximal third coefficient on the continuous pitch circle, i.e. $\{0.5, 4.5, 4.5\}$, subset of an augmented triad. The other trichord has opposite $a_3 = 2i - 1$, because[29] $\{6,7,11\}$ is closest to $\{6.5, 6.5, 11.5\}$ which is in perfect opposition with the other augmented triad position. To sum it up, both trichords have

---

[27] Actually this part is often analysed as a superposition of the tone-row and its retrograde. At first hearing however what is perceived is what I develop here.

[28] It is known that this division was very important for the composer, who strongly opposed easier fingerings proposed by his pianist Peter Stadlen. However, no less a pianist than Glenn Gould suppressed the high-risk hand-crossing at the beginning of the second half of variation 2.

[29] One could also use a symmetry argument.

a strong *augmented flavour*, but live in opposite directions of the harmonic spectrum in that respect, so that they neutralise each other, cf. Fig. 4.22.

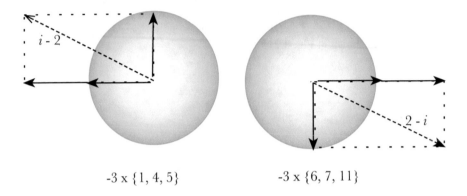

$$-3 \times \{1, 4, 5\} \qquad\qquad -3 \times \{6, 7, 11\}$$

Fig. 4.22. Coefficients a_3 for both trichords are opposite

The fourth coefficient is a simpler situation: $\{6,7,11\}$ is devoid of any minor-thirdish/octatonic flavour (it touches all three diminished sevenths) and the whole minor-thirdishness of the hexachord is supported by the first trichord, $\{1,4,5\}$. For all other coefficients, the trichords more or less combine their strengths into the hexachord's. The overall picture of this hexachord is highly chromatic, and somewhat whole-tonic (high values of first and sixth).

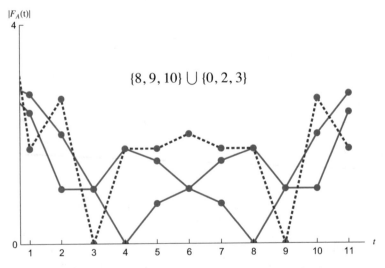

Fig. 4.23. Decomposition of the complement hexachord

It should come as no surprise that the second, complementary hexachord has the same Fourier distribution! But the decomposition into trichords introduces a slightly different fragrance: as we can see in Fig. 4.23, the (still opposite) third coefficients are much smaller ($\pm i$). Here it is a very chromatic trichord $\{8, 9, 10\}$ that is devoid of any specific diminished character (fourth coefficient nil), and the whole of this dimension in the hexachord is carried by the other trichord $\{0, 2, 3\}$.

All in all, this shows that despite the absence of isometry, the choice of contrasted constituent trichords enhances the balance between the hexachords, which goes well with Webern's use of symmetries in all three variations.

Another famous and much analyzed dodecaphonic example is the initial tone-row of Berg's *Lyrische Suite* op. 28 (Fig. 4.24). The following analysis adds a new perspective to the traditional analysis of hidden (fourth/fifth) cycles, like [71].

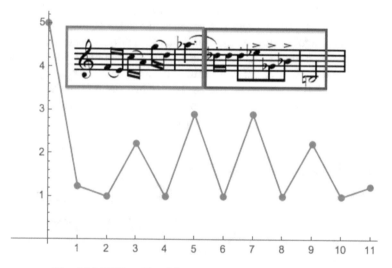

Fig. 4.24. DFT profile of the initial seven notes (or last five).

This time, there are conjoint rhythmic, melodic and dynamic reasons to segment this series into 7+5 notes, cutting between the high A♭ and the sequel. Of course, the DFT of FECAGDA♭ is identical (except the 0th coefficient) with the DFT of the remainder D♭E♭G♭B♭B, by Babbitt's *generalised* theorem. However the shape of Fig. 4.24 deserves commentary.

The large fifth (or seventh) coefficient is known as an indicator of diatonicity. Indeed, both parts are close to diatonic and pentatonic respectively (rotating the last B to the beginning of the series would allow a perfect decomposition of this sort). If we remove the A♭ (equivalently, we may decide to segment the phrase *before* the A♭, the two hexachords are isometric a tritone away), it yields the Guidonian hexachord, which has a neat DFT profile (see Fig. 8.26) with maximal diatonicity (it is saturated in fifths), and nil even coefficients, enhancing the contrast with the fifth content –

which is rhythmically obvious when one picks every other note, getting fifths FC, EA, CG, AD...

Another well-known feature of this tone-row which deserves further comment is the structure of *consecutive* intervals: though we know (from Babbitt's theorem again) that the overall intervallic structures of both parts are equal (up to a constant since their cardinalities differ here), the composer manages to pick up different consecutive intervals. Specifically, if interval $\delta = b - a$ appears between two consecutive notes in the first seven, the opposite interval appears in the sequel (e.g. FE vs. B♭B). This is a delicate construct to achieve by hand, and I leave it to the reader to construct the 48 'all-interval series' beginning with the seven white keys in some order (the last being precisely Berg's tone-row, up to a cyclic permutation).

I think that it is not so far-fetched to infer from this example one reason why Berg seems more amenable to untrained ears (in 20th century music): even in dodecaphonic music, he manages to keep a significant diatonic character. This idea is not original, but it can now be checked scientifically by using DFT. For instance, segmenting his sonata op. 1 every two seconds, the value of $|a_5|$ on each segment averages 1.57, a significantly large value. This should be researched more intensively of course[30], studying motives and especially hexachords throughout his work vs. Webern's and Schönberg's. I will venture just another (well-known) example of clear diatonicity in Berg, the initial and last bars of his Violin Concerto arpeggiating fifth cycles (as a four-note cycle and then a diatonic F major scale), and the main tone-row featuring remarkably diatonic hexachords, see Fig. 4.25.

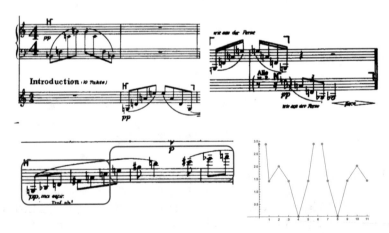

Fig. 4.25. First bars and tone-row in Berg's Violin Concerto, with its hexachords' clearly diatonic Fourier profile

---

[30] I carried out a cursory analysis of Berg's *Vier Stücke* op. 3, an 'intermediate' piece, atonal but not yet dodecaphonic; interestingly, it seems to exhibit much lower values of a_5.

In conclusion, DFT now provides precise, objective, quantitative measurements of diatonicity (or octatonicity, or whole-toneness, etc.) for almost any given piece of music.

4.3.3 Flat distributions

FLIDs

In his talk at the first MCM convention (Berlin 2007), Canadian theorist Jonathan Wild introduced FLIDs – Flat Interval Distribution Sets. The idea was a generalisation of the famous case of 'all-interval pc-set', e.g. for $A = \{0,1,3,7\} \subset \mathbb{Z}_{12}$, every interval occurs exactly once (except the tritone, because $7-1$ is the same as $1-7=6$ modulo 12).

Definition 4.33. $A \subset \mathbb{Z}_n$ is a FLID if $\mathrm{IC}(A)(k)$ is constant for $k = 1,2 \ldots n/2$.

Wild allowed the tritone interval $n/2$ when n is even, or else there are no possible FLIDs since a tritone must occur twice or not at all ($n/2 = -n/2$)[31]; we cannot take this view here because of Theorem 4.36 below and hence restrict FLIDs to odd values of n. One index which cannot be taken into account is 0, because $\mathrm{IC}(A)(0)$ is always the cardinality of A, larger than all other possible values of $\mathrm{IC}(k)$.

Actually the notion has been well studied in combinatorics under the name of 'difference sets'. There is a nice relationship with block designs[32]: if $D = \{d_1, \ldots d_k\} \subset \mathbb{Z}_n$ is such that any $b \neq 0$ in $\mathbb{Z}_n$ can be expressed in λ different ways as $d_i - d_j$, then the $D + \tau, \tau \in \mathbb{Z}_n$, form a (n,k,λ) block design.

Example 4.34. Let $n = 11$ and consider the *quadratic residues*, i.e. all squares reduced modulo n (for instance $5^2 \equiv 3 \mod 11$). Their set, $D = \{0,1,3,4,5,9\}$ is a 3-FLID: in $D - D$ all possible values (except 0) occur thrice, see Fig. 4.26 (this construction, known as *Hadamard difference sets*, works for prime powers $n \equiv 3 \mod 4$). The associated block design is $(11,6,3)$: any pair of translates of D, e.g. D and $D+3 = \{1,3,4,6,7,8\}$, intersects in exactly three points.

The last example is invariant under multiplication (squares of multiples are squares). More generally, since affine maps *permute* the values of the IC[33], we can state that

Proposition 4.35. *Any affine transform of a FLID $A \subset \mathbb{Z}_n$ (i.e. any $aA + b$ for a coprime with n) is also a FLID.*

---

[31] If the tritone is counted only once, then $\{0,1,3,7\}$ mod 12 or $\{0,2,3,5\}$ mod 6 (i.e. the French augmented sixth $A\flat CDF\sharp$ as a subset of a whole-tone scale) are FLIDs. A variant of Theorem 4.36 below could be established for this generalised definition, with the DFT's magnitude oscillating between two close values.

[32] A block design (n,k,s) is a collection of k-subsets of a n-set such that any pair of subsets shares s elements. When $s = 1, A$ is called a *projective plane*, like the famous Fano plane which is the reunion of the seven 'lines' $\{0,1,3\} + \tau$ in $\mathbb{Z}_7$ which intersect one another in one point exactly.

[33] Under the bijection $x \mapsto ax+b$, any interval δ is mapped to $a\delta$.

-	**0**	**1**	**3**	**4**	**5**	**9**
0	0	10	8	7	6	2
1	1	0	9	8	7	3
3	3	2	0	10	9	5
4	4	3	1	0	10	6
5	5	4	2	1	0	7
9	9	8	6	5	4	0

Fig. 4.26. Differences mod 11 of $D = \{0,1,3,4,5,9\}$.

Hence two such sets are usually considered equivalent if one is the affine image of the other.

Since affine maps also permute Fourier coefficients, this yields a neat proof of the easy implication of the following theorem, which links intervals and Fourier coefficient distributions:

Theorem 4.36. *A is a FLID iff its Fourier transform is flat. More precisely,*

$$IC(A) = (d,m,m,m\ldots) \iff |\mathscr{F}_A|^2 = (d+(n-1)m, d-m, d-m, \ldots d-m).$$

Remark 4.37. By a continuity argument, this means that the dispersion of values of the DFT (the 0th coefficient excepted) is correlated to the dispersion of the intervallic distribution: both are nil for FLIDs. We have studied the opposite case before: maximum values for one Fourier coefficient coincide with maximum occurrences for a given interval. Explicit but messy formulas for these dispersions can be computed.

Proof. The direct implication is straightforward, since $\widehat{IC(A)} = |\mathscr{F}_A|^2$: for $IC(A) = (d,m,m,m\ldots)$ one computes its Fourier transform,

$$|\mathscr{F}_A|^2(k) = d + \sum_{t=1}^{n-1} me^{-2i\pi kt/n} = (d-m) + \sum_{t=0}^{n-1} me^{-2i\pi kt/n} = d-m \text{ for } k \geqslant 1.$$

The value in 0 is the 'cardinality' of $IC(A)$, i.e. the sum of its elements $d+(n-1)m$.

The reverse implication is trickier. My original proof in [13] uses the algebra of circulating matrixes isomorphic with Fourier space. Here is a shorter one with DFT only, but it is not constructive.

Assume that $|\mathscr{F}_A|^2$ is flat, i.e. $|\mathscr{F}_A|^2 = (k,\ell,\ell,\ell\ldots)$ for some $k,\ell \in \mathbb{R}_+$. Define $d,m \in \mathbb{R}$ such that $d-m = \ell, d+(n-1)m = k$; then by the direct computation, the Fourier transform of the distribution $f = (d,m,\ldots m)$ is $|\mathscr{F}_A|^2$. Since DFT is bijective, and $\widehat{IC(A)} = |\mathscr{F}_A|^2 = \hat{f}$, we have $IC(A) = f = (d, m, \ldots m)$, i.e. $IC(A)$ is flat.

Large determinants

An equivalent characterisation stems from the following remark. The determinant of the circulating matrix associated with A (see Section 1.2.3) is simply the product of its Fourier coefficients: $\det(\mathscr{A}) = \prod_k \mathscr{F}_A(k)$. Consider $|\det(\mathscr{A})|^2$, which is the product of the Fourier coefficients of $IC(A)$. From Parseval-Plancherel's identity (Theorem 1.8),

$$\sum_{k \in \mathbb{Z}_n} |\mathscr{F}_A(k)|^2 = nd \text{ where } d = \#A.$$

As we have stated again, $\mathscr{F}_A(0) = \#A$ cannot vary. But in order to maximise the product of the other Fourier coefficients $\prod_{k=1}^{n-1} |\mathscr{F}_A(k)|^2$ under the condition $\sum_{k=1}^{n-1} |\mathscr{F}_A(k)|^2 = (n-d)d$, one must have them all equal.[34] Hence

Proposition 4.38. *Among all d-subsets $A \in \mathbb{Z}_n$, the maximal possible value of $|\det(\mathscr{A})|$ is reached when A is a FLID.*

Geometrically, this means that the columns of $\mathscr{A}$ are the least colinear as possible, i.e. that the translates $A, A+1, A+2 \ldots$ are as much apart (in $\mathbb{R}^n$) as possible i.e. that their mutual angles are as close to a square angle as possible.

FLIDs do not exist for any pair (n, d)[35], but this yields an explicit universal majoration:

$$\#A = d \Rightarrow |\det(\mathscr{A})| \leqslant d \left(d \frac{n-d}{n-1} \right)^{\frac{n-1}{2}}.$$

For instance, for 4-subsets of $\mathbb{Z}_{12}$, the maximum determinant is reached for the all-intervals tetrachords $\{0, 1, 3, 7\}$ (or $\{0, 1, 4, 6\}$) and is equal to 1,024, though the formula's upper bound yields about 1,421; there are no genuine FLIDs in $\mathbb{Z}_{12}$ because of the tritone doubling.

Perhaps this notion of the size of the determinant should warrant additional research. Obviously it is

- nil for subsets which tile;
- small for subsets with irregular interval distribution, like ME sets;
- and maximal for FLIDs.

FLIDs which tile

When the multiplicity m of all intervals in a FLID A is equal to 1, we reach a very interesting situation, because A tiles almost all of $\mathbb{Z}_n$:

---

[34] This is well known and can be proved for instance with convexity arguments.
[35] At least if one insists on actual pc-sets, i.e. distributions with values in 0-1.

Definition 4.39. *A Golomb ruler is a set A such that all difference values occur exactly once:*

$$a_i - a_j = a_k - a_l \iff (i,j) = (k,l)$$

It is perfect *if all possible values (except 0) are obtained once, i.e.*

$$\mathrm{IC}(A) = (d, 1, 1, 1 \ldots).$$

A Sidon set is a set A such that all sum values occur exactly once:

$$a_i + a_j = a_k + a_l \iff (i,j) = (k,l)$$

It is complete *if all possible values (except 0) are obtained once, i.e. A tiles a subset of $\mathbb{Z}_n$.*

Hence a perfect Golomb ruler in $\mathbb{Z}_n$ is a 1-FLID.

Proposition 4.40. *Sidon sets = Golomb rulers.*

Proof. $a_i - a_j = a_k - a_l$ has a unique solution $\iff a_i + a_l = a_k + a_j$ has a unique solution. $\qquad \square$

This trivial proposition yields a very nice link between intervallic studies and tilings; unfortunately there is no way these sets can provide true tilings of the whole of $\mathbb{Z}_n$. For instance $\{0,1,3\}$ only tiles $\{0,1,2,3,4,6\}$ in $\mathbb{Z}_7$. Even almost FLIDs like $\{0,1,4,6\}$ in $\mathbb{Z}_{12}$ cannot tile without overlapping[36] since

Proposition 4.41. *An all-interval set intersects any of its translates. The cardinality of the intersection $A \cap (A+t)$ is $\mathrm{IC}(A)(t)$.*

Though difference sets are mostly studied in $\mathbb{Z}$ (or even larger structures) they deserve a mention in this book.[37] For one thing, Sidon originally created the eponym sets during his investigation of Fourier series.[38] Some very specific constructions are known which yield spectacular results.

For instance, in [80] Singer inadvertently constructed a superb 1-FLID[39], alias Sidon set:

Theorem 4.42. *For any prime p there exists a subset A of $\mathbb{Z}_n$ with $p + 1$ elements, where $n = p^2 + p + 1$, such that the intervallic distribution is uniform: $\mathrm{IC}(A)(k) = 1$ for all k (except $k = 0$ of course).*

---

[36] Composer Tom Johnson has practiced with graphs between pc-sets with the relationship 'not intersect', see for instance [52].

[37] See also the notion of spectral set which can be expressed in terms of differences, cf. Proposition 3.58.

[38] Sidon sets are still instrumental in the study of lacunar and/or random Fourier series in Harmonic Analysis.

[39] The construction also yields $\left(\dfrac{p^{n+2} - 1}{p - 1}, \dfrac{p^{n+2} - 1}{p - 1}, \dfrac{p^{n+2} - 1}{p - 1} \right)$ difference sets, cf. [29].

The construction is non-trivial, making use of cubic extensions of finite fields (which appear to crop up often, quite unexpectedly, in tiling theory, see [6] for instance). Examples are $\{0,1,4,6\}$ for $p=3, n=13$ or $\{0,1,4,6,13,21\}$ for $p=5, n=31$ or $\{0,1,6,15,22,26,45,55\}$ for $p=7, n=57$.

It is easy to check [29] that for these distributions

Proposition 4.43.

$$\forall k \neq 0 \quad \mathscr{F}_A(k) = \sqrt{p}.$$

Proof. This a special case of Theorem 4.36 which also yields the reciprocal. In this case, one can compute directly

$$|\mathscr{F}_A(k)|^2 = \mathscr{F}_A(k)\overline{\mathscr{F}_A(k)} = \sum_{x,y \in A} e^{-2i\pi k(x-y)/n}$$

$$= \sum_{x,y \in A, x \neq y} e^{-2i\pi k(x-y)/n} + \sum_{x \in A} e^{-2i\pi k(x-x)/n}$$

$$= \sum_{z=1}^{n-1} e^{-2i\pi kz/n} + \sum_{x \in A} e^0 = -1 + (p+1) = p.$$

For practical purposes, it is often convenient to assume that the Singer set begins with $(0,1)$ (up to affine transform). A feature these sets share with FLIDs is the stability of their class under affine transformations, since these transformations only permute the interval distribution. Jon Wild sent me the following collection of FLIDs/Singer sets in $\mathbb{Z}_{31}$:

$$(0,1,3,8,12,18), (0,1,4,10,12,17), (0,1,16,18,22,29),$$
$$(0,1,11,19,26,28), (0,1,15,19,21,24)$$

which are all affine images one of another[40] and can be arranged to tile $\mathbb{Z}_{31}^*$ with appropriate translations.[41] This is an instance of different but homometric tiles which have perfectly balanced saliency for all coefficients. It might seem strange that the tiles have no nil Fourier coefficients in this situation. But it could be surmised from the fact that they tile $\mathbb{Z}_n \setminus \{0\}$, complement of the Dirac distribution (neutral element for $*$), whose DFT is non singular (it is $(n-1, -1, -1, -1, -1, \ldots)$).

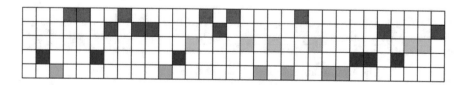

Fig. 4.27. Tiling with different Singer sets modulo 31

---

[40] See in exercises.

[41] He also found tilings with two or four tiles out of these five.

The interplay with the affine group suggests looking for stability features. Quite often, a Singer set A (or a FLID, actually) is invariant under an affine map or simply multiplication by a constant, i.e. $pA = A$.

Definition 4.44. *Such a p is called a multiplier of A. The set of all multipliers of A is a subgroup of $\mathbb{Z}_n^*$.*

For instance $(1,2,4)$ in $\mathbb{Z}_7$ has multiplier 2: it is actually the orbit of 1 under multiplication by 2. In the above example, if $A = (0,1,4,6,13,21)$ we can see that $5A = (0,5,20,30,65 = 3,105 = 12)$, i.e. $5A + 1 = (0,1,4,6,13,21) = A$ or $5(A+8) = A+8$, i.e. 5 is a multiplier of $A+8 = (8,9,12,14,21,29) = (1,5,25) \otimes (8,12)$. It is conjectured that in general, some *translate of* a FLID has multipliers.[42] This might be interesting for composers who play with affine transforms, and should perhaps warrant exploration with non-commutative Fourier transform in the affine group.

To round up this discussion and generalise the last example, let us mention that other tilings of $\mathbb{Z}_n \setminus \{0\}$ by augmentation have been discovered in investigating composer Tom Johnson's autosimilar melodies [54]. To quote him, the absence of 0 is a welcome respite – he devotes a whole chapter to 'punching some holes in the melody,' because:

> The musical interest can be quite a bit greater after punching some holes, however. The different durations define themes with more character, that can be more easily recognised, and this is a great advantage when we are trying to hear the theme in two or three different tempos.

Example 4.45. Consider motif $A = \{1,2,4\}$. It is an orbit of $x \mapsto 2x$ in $\mathbb{Z}_7$. The other orbit is $\{3,6,12 = 5\}$. Thus $A \cup 3A = \{1,2,3,4,5,6\}$. From there, one can associate one note to each orbit and thus reach a melody with ratio 2 autosimilarity (picking every other note yields the same melody, twice slower); or tile $\mathbb{Z}_7^*$ with a cross-section of the orbits – say $S = \{2,3\}$ – and its augmentations $\{4,6\}$ and $\{1,5\}$, see Fig. 4.28 with the autosimilar melody first, then the tiling by augmentation.

Fig. 4.28. Autosimilar melody and dual tiling by augmentation

For a larger example, take $n = 31, A = \{1,2,4,8,16\}$.
The other orbits, $3A = \{3,6,12,17,24\}, 5A = \{5,9,10,18,20\}, 7A = \{7,14,19,25,28\}$,

---

[42] At least when p divides $n - m$. This is equivalent to invariance under (some kind of) affine transformation.

$11A = \{11, 13, 21, 22, 26\}$ and $15A = \{15, 23, 27, 29, 30\}$, partition $\mathbb{Z}_{31}^*$. Now choose any cross-section of these orbits, say $S = \{1, 3, 5, 7, 11, 15\}$; another partition uses $S, 2S, 4S, 8S, 16S$. This enables one to construct a tiling by binary augmentation, see Fig. 4.29, reminding of Fig. 4.27.

Fig. 4.29. Tiling by augmentation modulo 31

Finally, there are tilings of the *complement of a subgroup* using all different ratios of augmentation exactly once: in [53], T. Johnson cites a tiling with augmentations of $\{0, 1, 3\}$ with ratios 1,2,3,4,5,6,7,8 which leave aside all beats congruent with 2 mod. 3, see Fig. 4.30. More complicated examples are also mentioned but we are straying away from the topic of this book. See [4], online supplementary, for details and generalisations.

An open question is the characterisation of these objects by way of using Fourier transform of the multiplicative abelian group $\mathbb{Z}_n^*$, whose structure varies a lot according to the value of n and boils down to multidimensional DFT because of the decomposition of any abelian finite group into a product of cyclic groups.

for Jon Wild
Extra Perfect

Tom Johnson

17/09/2008

Fig. 4.30. Tiling with 013 and augmentations, leaving holes every third eighth-note

Heisenberg's uncertainty principle

We have now seen enough varied material, from both ends of the spectrum so to speak, that we can perhaps address some broader issues. I will broach the question of the cardinality of A (i.e. pcs, for scales or chords, or beats, for periodic rhythms) vs. the zeroes of $\mathscr{F}_A$ which have been so important in different situations.

Much of the material in this book addresses questions of retrieval (phase retrieval for homometry, support retrieval for a complement of a tiling motif, and so on). There is a definite advantage then, when the number of coefficients to retrieve is small. This can be foretold in some measure. It is well known, at least informally, that the DFT spreads when the information (size of time window, sampling, smaller periodicity...) decreases. A few examples:

- If A has a period $d \mid n$, A has at least n/d elements and $\mathscr{F}_A$ is nil except on the subgroup with d elements.
- If A is a FLID (see 4.3.3) then $\#A$ may be small but $\mathscr{F}_A$ never vanishes.
- If A tiles $\mathbb{Z}_n$, then $\#A$ divides n, i.e. is usually comparatively small; $\mathscr{F}_A$ vanishes on $Z(A)$, the union of all elements whose order belongs to $R(A)$ (a reunion of orbits of the action of $\mathbb{Z}_n^*$, cf. Theorem 3.11) and hence a sizeable subset of $\mathbb{Z}_n$. Besides, if A tiles with B, then $\#Z(A) + \#Z(B) \geqslant n - 1$ while $\#A \times \#B = n$.

These relationships between the zeroes of $\mathscr{F}_A$ and those of $\mathbf{1}_A$ can be quantised by the following result, commonly used by researchers in various fields but not really pointed out in textbooks, and reminiscent of Heisenberg's famous inequality in quantum physics:

Theorem 4.46 (Discrete Uncertainty Principle).
Let f be a distribution on $\mathbb{C}^n$, $\widehat{f}$ its DFT, and let $\mathrm{Supp}(f)$ stand for $\{x \in \mathbb{Z}_n \mid f(x) \neq 0\}$. Then

$$\#\,\mathrm{Supp}(f) \times \#\,\mathrm{Supp}(\widehat{f}) \geqslant n$$

This means that if $\widehat{f}$ has few zeroes then f has many, and conversely. Notice that $\mathrm{Supp}(\widehat{f})$ is just the complement of the zero set of the Fourier transform.

Proof. Recall Parseval-Plancherel equality (Theorem 1.8):

$$\sum |\widehat{f}(k)|^2 = n \sum |f(k)|^2,$$

and plug in the following elementary inequalities:

$$\sup_x |f(x)| = \sup \left| \frac{1}{n} \sum_k \widehat{f}(k) e^{2i\pi kx/n} \right| \leqslant \frac{1}{n} \sum_k |\widehat{f}(k)| \quad \text{(inverse DFT)}$$

$$\sum |f(k)|^2 = \sum_{k \in \mathrm{Supp}(f)} |f(k)|^2 \leqslant \#\,\mathrm{Supp}(f) \times \sup |f(x)|^2$$

$$\sum |\widehat{f}(k)| = \sum_{k \in \mathrm{Supp}(\widehat{f})} 1 \times |\widehat{f}(k)|$$

$$\leqslant \sqrt{\sum_{k \in \mathrm{Supp}(\widehat{f})} 1^2} \sqrt{\sum_{k \in \mathrm{Supp}(\widehat{f})} |\widehat{f}(k)|^2} = \sqrt{\#\,\mathrm{Supp}(\widehat{f})} \sqrt{\sum |\widehat{f}(k)|^2},$$

this last one being Cauchy-Schwarz inequality. Combining all this,

$$\sum |\widehat{f}(k)|^2 = n \sum |f(k)|^2 \leqslant n \# \operatorname{Supp}(f) \sup |f(x)|^2 \leqslant n \# \operatorname{Supp}(f) \left(\frac{1}{n} \sum_k |\widehat{f}(k)| \right)^2$$

$$\leqslant \frac{1}{n} \# \operatorname{Supp}(f) \# \operatorname{Supp}(\widehat{f}) \sum |\widehat{f}(k)|^2,$$

hence the result.

The inequality is sharp: if n is a square $n = d^2$ then for $A = \{0, d, 2d, \ldots d^2 - d\} = d\mathbb{Z}_n$, the DFT $\mathscr{F}_A$ is proportional to $\mathbf{1}_A$ itself and both supports have d elements.

Improvements of this lower bound are known in the case of very simple cyclic groups (for instance when n is prime) – see [83] from which I borrowed much of this section – but are so far of little interest to musicians.

It should be noted that the extreme cases of maximal vs. nil Fourier coefficients are by no means contradictory. It could even be argued that an ubiquitous motif like CDE ("Brother John"...) owes much of its versatility to the dual facts that on the one hand it tiles, having several nil Fourier coefficients ($a_2, a_4 \ldots$), but on the other hand it exhibits strong characters: a_6 is maximal since CDE is a chunk of whole-tone scale, cf. Fig. 8.8.

Exercises

Exercise 4.47. Generated scales: peruse the online catalog for Fourier profiles of generated scales in $\mathbb{Z}_{12}$, on

> `http://canonsrythmiques.free.fr/MaRecherche/photos-2/.`

Exercise 4.48. Saturation: find musical instances of pc-sets saturated in major thirds, like $\{0, 1, 4, 8\}$ or $\{0, 1, 4, 5, 8\}$.

Exercise 4.49. Generated scales: create a scale with 20 generators in some $\mathbb{Z}_n$.

Exercise 4.50. Generated scales: find other occurrences of the complement set of the tresilo rhythm in tango or elsewhere.

Exercise 4.51. ME sets: compute instances of $\operatorname{ME}_{(11,7)}, \operatorname{ME}_{(19,7)}, \operatorname{ME}_{(24,7)}$ ('diatonic scales' in other divisions of the octave).

Exercise 4.52. ME sets: find some other type III ME sets.

Exercise 4.53. Saliency: using the online catalog of Fourier profiles, study the saliencies of some pc-sets in a musical piece of your own choosing. Early 20th century is a good starting point.

Exercise 4.54. FLID: find some FLID, for instance the Hadamard kind for $n = 23$ or $n = 43$, and check its interval content.

Exercise 4.55. FLID: prove Proposition 4.41.

Exercise 4.56. Singer sets: find the affine maps transforming the first motif in Fig. 4.27 in each of the others. Check that they are indeed Singer sets.

Exercise 4.57. Pick up a tiling $A \oplus B = \mathbb{Z}_n$ in the examples given or otherwise. Check Heisenberg's uncertainty principle on each factor.

5

Continuous Spaces, Continuous FT

Summary. The formula for the Fourier Transform $\widehat{f}(t) = \sum f(k)e^{-2ik\pi t/n}$ can be extended to continuous settings in several ways: transforming the discrete $\sum$ into a continuous $\int$ with some appropriate (usually Lebesgue) measure[1], i.e. summing on an infinite set; or having the variable t move on the real line instead of a cyclic group; or having the frequency $\frac{2\pi}{n}$ become infinitesimal, perhaps keeping value $\frac{2k\pi}{n}$ constant while both k and n grow to infinite. A last variant considers ordered collections of pcs with fixed cardinality, leading unexpectedly to a good measure of quality for temperaments such as might have been used by J.S. Bach. All these changes, advertised by various researchers [25, 88], require however precise definitions of their contexts and limitations, which will be scrupulously enunciated hereafter. Several practical situations of music devised by playing directly in some continuous Fourier space have occurred in recent years, and are reviewed in the last section.

5.1 Getting continuous

The model of pc-sets which has been hitherto followed in this book presupposes a fixed division in n parts of the octave (or a quantisation of the timeline); the domain of the Fourier transform is then $\mathbb{C}^n \approx \mathbb{C}^{\mathbb{Z}_n}$, the vector space of distributions with dimension n. The components of the vectors in this space are quantities of a given pc, e.g. 100% of C, 0% of C♯ and so on. In this space there is no such thing as C-quartertone. However, the introduction of orbifolds[2] by Callender, Quinn and Tymoczko in 2005 for the purveying of good spaces for voice leadings [26] led to the consideration of Fourier transform with continuous pitch-class values. One could argue that this topic is hence beyond the purpose of the present book. Nonetheless, I feel it necessary to quickly sketch what this other DFT is about, if only to clarify the distinction between the different kinds; besides, since it addresses a finite number of pitches or pitch-classes, it is also discrete, in a manner of speaking. I will accordingly

---

[1] See however [2] where for instance the Z-relation is defined using a Haar measure on a compact topological group.

[2] The first mention of this notion by Guerino Mazzola in a music-theoretical context in [66], in German, went largely unnoticed at the time.

© Springer International Publishing Switzerland 2016
E. Amiot, *Music Through Fourier Space*, Computational Music Science,
DOI 10.1007/978-3-319-45581-5_5

provide rigorous mathematical definitions which the aforementioned authors had to shun in their publications, aimed at less mathematically tolerant readerships than the present opus.

Approach through real frequencies

The formula that gets generalised is actually the less general one, for a DFT of a pc-set (instead of a distribution):

$$\mathscr{F}_A(t) = \sum_{k \in A} e^{-2ik\pi t/n}.$$

At this point we are down to much less than a vector space: a collection of 2^n maps from $\mathbb{Z}_n$ to $\mathbb{C}$ (or equivalently 2^n vectors in $\mathbb{C}^n$). But in this formula, more items than A can be considered variable. Callender starts his highly readable [25] by considering $1/n = \ell$, the frequency, as a real variable, along with k and t. This is the highest possible degree of generality, and all subsequent definitions can be derived from it by restriction. Formally, we should consider the set $\mathscr{P}_0(\mathbb{R})$ of finite subsets of $\mathbb{R}$, and define

$$\begin{array}{cccc}
\mathscr{C}: \mathscr{P}_0(\mathbb{R}) \times \mathbb{R} \times \,]0,+\infty[& \longrightarrow & \mathbb{C} \\
(A, \quad t, \quad \ell) & \longmapsto & \sum_{k \in A} e^{-2ik\pi t\ell} \, .
\end{array}$$

Callender first shows, for a fixed A, the influence of the frequency parameter ℓ and how this map that I will still call $\mathscr{F}_A : \mathscr{C}(A,.,\ell)$ defines the *coincidence function* of pc-set A with a ℓ-cycle. Let us reproduce[3] his example $A = \{0,5,15\} \subset \mathbb{Z}$ (not $\mathbb{Z}_{12}$) with the graph of $|\mathscr{F}_A(x)|$ in Fig. 5.1.

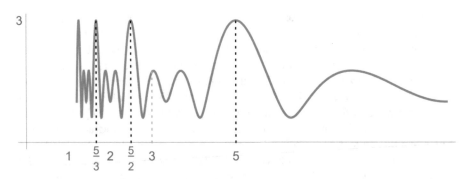

Fig. 5.1. Magnitude of $\mathscr{F}_A(1/x)$ for $A = \{0,5,15\}$

The maximums for $x = 5/n, n \in \mathbb{N}^*$ are immediately apparent. The author convincingly explains the secondary maximums (for $x = 3$) by the closeness of $A = \{0,5,15\}$ to $A' = \{0,\mathbf{6},15\}$ which does admit an absolute maximum for $x = 3$, see Fig. 5.2.

---

[3] I am grateful for his permission.

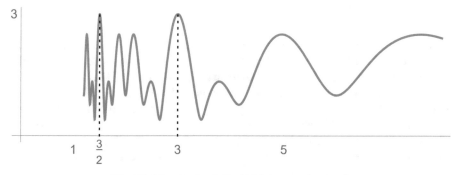

Fig. 5.2. Magnitude of $\mathscr{F}_{A'}(1/x)$ for $A = \{0, 6, 15\}$

More generally, it will come as no surprise to the readers who made it this far in this book that $|\mathscr{F}_A(1/\ell)| = d = \#A$ if and only if A is a subset of an arithmetic sequence with period ℓ (or equivalently, frequency $1/\ell$), e.g. the maximum is reached for $\ell = 5$ and its divisors $5/n$ for set A above (resp. $\ell = 3$ for set A'): it is a consequence of lemma 4.23 in a much more liberal context, wherein one is at liberty to have all the exponential terms in the sum be equal. When A is *not* a subset of a regular progression, the theoretical maximum $\#A$ cannot be reached, but it is worthwhile to notice that one may get as close to it as desired, because of the following diophantine approximation result:

Proposition 5.1. *For any (multi)set $A \subset \mathbb{Z}$ with cardinality d, $|\mathscr{F}_A(x)|$ can be rendered as close to d as desired.*

Proof. This follows from a classical number theory result:

Lemma 5.2 (Dirichlet's simultaneous rational approximation theorem). *For any set of real numbers $\{\alpha_1, \ldots, \alpha_d\}$ and integer $N > 0$, there exist integers $p_1 \ldots p_d$ and $1 \leqslant q \leqslant N$ such that*

$$\forall i = 1 \ldots d \quad \left| \alpha_i - \frac{p_i}{q} \right| \leqslant \frac{1}{q N^{1/d}},$$

which can be found in textbooks in number theory or online (it can be proved using the pigeonhole principle).

Now set $k_i/\ell = \alpha_i$ when k_i runs through A and ℓ is some large integer. For $N \geqslant \ell$ we get from the above lemma

$$\left| \frac{q k_i}{\ell} - p_i \right| \leqslant \frac{1}{N^{1/d}};$$

since this majoration is arbitrary small, $e^{-2i\pi q k_i/\ell}$ is arbitrarily close to $e^{-2i\pi p_i} = 1$ by continuity of the exponential map.[4] Hence $\mathscr{F}_A(q/\ell)$ is arbitrarily close to $1 + 1 + \ldots 1 = d$, as announced.

This result is well known in standard spectrum analysis.[5] It can be paraphrased as follows:

> Any finite spectrum is approximately harmonic if the (virtual) fundamental is taken at a low enough frequency.

---

[4] Independently of the index i, which is the main point of the proof.

[5] See the seminal [85] which explores the perception of virtual pitches.

DFT on an orbifold

Let us return to the continuous frequency FT and aim towards a DFT of elements of orbifolds. In order for the formula to make sense for pitch-*class* sets (say modulo 12), the frequency has to be taken equal to an integer multiple of $1/12$: $\ell \in \frac{1}{12}[\![1,12]\!]$. At this moment (and none other) we cross again the definition that we had used throughout this book. The next step is substantial: having decided on a fixed division of the octave (say in 12 parts, i.e. 12 pcs), and a finite number of relevant values of the index $t = 0\ldots 11$, the orbifolds' proponents allow the variables to wander away from integer values. But even though the philosophy may be debatable, the following mathematical definition is sound[6]:

Definition 5.3. *For any $k \in \mathbb{N}^*$, the map*

$$\mathscr{F}_k : \left(\mathbb{R}/(12\mathbb{Z})\right)^k \times \mathbb{N} \to \mathbb{C}$$
$$(A \quad , t) \mapsto \sum_{a \in A} e^{-2ia\pi t/12}$$

is well defined, from the set of k-sets of continuous pcs modulo 12 times the integers, to the complex numbers. Moreover, $|\mathscr{F}_k(A,t)|$ does not change when A is transposed, inversed, or even permuted: this induces a well-defined map $|\mathscr{F}_k|$ on the orbifold of k-(multi)sets quotiented by any or all of the above operations.

This means that the beautiful, smooth pictures of *continuous* landscapes of pc-sets such as Figs. 4.1 or 5.3 are not devoid of meaning. *However, it is important to be aware that they are not the 'mindscapes of chords' that Quinn had originally in mind, and which took place in a discrete universe.* Moreover, this $|\mathscr{F}_k|$ is confined to k-chords, contrariwise to Quinn and Lewin's original DFT of pc-(multi)sets.

Approach through quantisation

Still, there is an interesting connection between the discrete case and the orbifolds: both Callender and Tymoczko [25, 88] put forth a *quantisation* of the continuous pc-circle. Best is to quote the former's example of four pc-sets with decimal values,

$$P = \{0, 0.46, 0.95, 1.41\} \; Q = \{0, 1.01, 6, 7.01\} \; R = \{0, 3, 6, 9\}$$
$$R' = \{0, 3.005, 6, 9.005\},$$

and the two first Fourier coefficients are

$$\mathscr{F}_4(P, 1) \approx 3.6 - 1.39\,\mathrm{i}; \; \mathscr{F}_4(Q, 1) = 0; \qquad \mathscr{F}_4(R, 1) = 0; \; \mathscr{F}_4(R', 1) = 0;$$
$$\mathscr{F}_4(P, 2) \approx 2.5 - 2.3\,\mathrm{i}; \quad \mathscr{F}_4(Q, 2) \approx 3.0 - 1.7\,\mathrm{i}; \; \mathscr{F}_4(R, 2) = 0; \; \mathscr{F}_4(R', 2) \approx 0.$$

---

[6] Notice however that the parameter $t \in \mathbb{N}$ must be an integer, not an integer class for the definition to work, since even modulo 12, kt will be different from $k(t + 12)$ in general. The equivalence modulo octave of intervals must perforce be lost in the process. So is the linearity of DFT, since the ambient space is only locally linear; also lost is its invertibility: there can be no inverse DFT in this context.

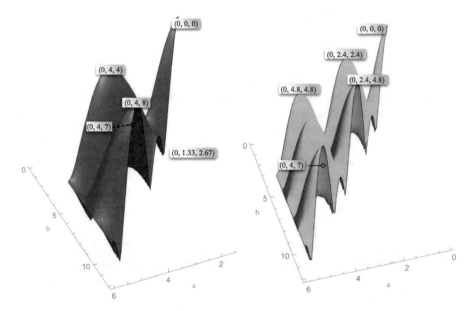

Fig. 5.3. Magnitude of $|a_3|$ and $|a_5|$ for 3-chords $(0, a, b)$

I modified Callender's notations for consistency with the above so that we can easily discriminate between P (unbalanced) and the others, or (P, Q) (irregular) and (R, R') (close to regular polygon). But in order to differentiate between R and R', one needs to 'jump far ahead at the 1200th harmonic', i.e. compute:

$$\mathscr{F}_4(P, 1200) \approx 4 = \mathscr{F}_4(R, 1) \approx \mathscr{F}_4(Q, 1) \text{ while } \mathscr{F}_4(R', 1) = 0.$$

In dividing each semitone by 200, we managed to place all pc-sets as subsets of regular polygons, i.e. they have integer values in a 1200-Temperament! The approximation results mentioned above allow us to generalise this example: *with a high enough resolution it is always possible to discriminate between distinct pc-sets.*[7] Though the price to pay may be too high if [25]

> ...the spectrum of a given pc-set is [...] within an infinite dimensional Fourier space....

Another interesting way to look at this quantisation is more developed in [88], where Tymoczko puts forth a convincing heuristic argument: any *rational k* turns into an integer for a large enough n, e.g. $k = 15/7$ is not an integer modulo 12, but $7k = 15$ modulo $7 \times 12 = 84$ is. More formally, one can recall the oversampling of distributions as described in the first chapter and write

---

[7] And even to single out a determinate subcollection from a sequence of pc-sets, like P, Q, R in the last case.

$$\sum f(k)e^{-2i\pi kt/n} = \sum f_p(pk)e^{-2i\pi pkt/(pn)}$$

$$\text{where } f_p(x \pmod{pn}) = f(\frac{x}{p} \pmod{n}).$$

In this way, the continuous DFT appears as a natural extension of the discrete temperament case. This goes well so long as rational numbers are used, since the value for the DFT does not change when all k's are turned into integers by multiplying the modulo, n, by larger and larger factors p, i.e. dividing the octave into finer and finer parts. It can (and indeed should) be argued that even when $k \notin \mathbb{Q}$, the term $f_p(pk)e^{-2i\pi pkt/(pn)}$ admits a limit when $p \to +\infty$, enabling rigourous definition of the FT of a (multi)set with arbitrary elements. I am still worried, though, by the jump from finite structures to the continuous circle of pitch-classes and especially the change of topology.[8]

While the progressive approach by quantisation is quite worthy of interest in itself, it should not be confused with the known properties of the DFT of distributions, which preserve the clear advantage of allowing comparison of pc-sets with variable cardinality.[9] Also, this process of finer divisions suggests that the natural set of values for pitch-classes is the field $\mathbb{Q}$ of rational numbers, which hints that this DFT would be nicely suited to studying (just) temperaments. This was actually done with yet another definition of DFT, as we will see in the next section.

5.2 A DFT for *ordered* collections of pcs on the continuous circle

Thomas Noll suggested in 2005 another DFT for *ordered sequences* of complex numbers with unit length, modeling pitch classes modulo octave, more suited to *scales* than to *chords*:

Definition 5.4. *Let any note modulo octave be given by a real number between 0 and 1; this means choosing a reference note (say C) and measuring all intervals from there in cents/1200, or alternatively[10] taking the dyadic logarithm of the frequency, modulo 1: $f \mapsto \dfrac{\ln f}{\ln 2} \pmod{1}$. Then the DFT of the ordered scale $\mathscr{A} = (a_1, a_2, \ldots a_n)$ where $a_1, \ldots a_n \in [0, 1[$ is the map[11]*

---

[8] In a parallel case, when [21] develops the first hexachordal theorem on the continuous circle as a limit of the $\mathbb{Z}_n$ case when n goes towards the infinity, it is not obvious *why* the theorem will only stand for measurable subsets of the circle: the authors proceed by analogy and only mention in a footnote that 'for instance, Lebesgue-integrable suffices.' The proper approach is by way of the Haar measure on a compact group, as developed in [2].

[9] Chapter 6 deals with another continuous model – a torus – where discrete pc-sets with all cardinalities coexist.

[10] See the algorithm in Section 3.3 with the former definition.

[11] Notice the changed notation – this is the third distinct definition of a Fourier transform of a collection of pcs.

$$\mathfrak{S}_{\mathscr{A}} : t \mapsto \frac{1}{n} \sum_{k=1}^{n} e^{2i\pi a_k} e^{-2i\pi kt/n} = \frac{1}{n} \sum_{k=1}^{n} e^{2i\pi \left(a_k - kt/n\right)}$$

where t is defined modulo n (it is the Fourier transform of the map $k \mapsto e^{2i\pi a_k}$ from $\mathbb{Z}_n$ to $S^1 \subset \mathbb{C}$).

The values $\mathfrak{S}_{\mathscr{A}}(0), \mathfrak{S}_{\mathscr{A}}(1) \dots \mathfrak{S}_{\mathscr{A}}(n-1)$ are the Fourier coefficients of scale $\mathscr{A}$.

For instance, the (equal-)tempered C major scale in step order would be $C_M = (0, 1/6, 1/3, 5/12, 7/12, 3/4, 11/12)$ and its DFT is

$$\mathfrak{S}_{C_M} : t \mapsto \frac{1}{7}\left(1 + e^{-\frac{2i\pi t}{7}} + e^{-\frac{4i\pi t}{7}} + e^{-\frac{6i\pi t}{7}} + e^{-\frac{8i\pi t}{7}} + e^{-\frac{10i\pi t}{7}} + e^{-\frac{12i\pi t}{7}}\right).$$

I stress the point that the ordering of the pcs is specified: *permuting the pcs changes the scale*, which is a sequence, not a set. This new DFT exhibits interesting geometrical features:

Proposition 5.5 (Noll 2006).
  If $\mathscr{A}$ is a generated scale, then its Fourier coefficients are aligned.

Proof. Assume $\forall k, a_k = k\theta$ (mod 1) for some generator θ (we take the scale in generation order, not step order). Then

$$n \times \mathfrak{S}_{\mathscr{A}}(t) = \sum_{k=1}^{n} e^{2i\pi k(\theta - t/n)} = \frac{e^{2i\pi(\theta - t/n)} - e^{2i\pi n(\theta - t/n)}}{1 - e^{2i\pi(\theta - t/n)}} = \frac{e^{2i\pi(\theta - t/n)} - e^{2i\pi n\theta}}{1 - e^{2i\pi(\theta - t/n)}},$$

since $e^{-2i\pi t} = 1$. Now this expression is homographic in $\xi_t = e^{2i\pi(\theta - t/n)}$, which moves on the unit circle (actually on a regular polygon):

$$n \times \mathfrak{S}_{\mathscr{A}}(t) = \psi(\xi_t) = \frac{\xi_t - e^{2i\pi\theta}}{1 - \xi_t}.$$

A homography maps a circle onto a circle or a straight line. This is the latter case and not the former, since when $\xi_t \to 1$ the expression gets infinite.

It is worth noticing that the result stands for a WF scale in step order, since the step order can be deduced from generating order by multiplication, i.e. changing t. In Fig. 5.4 we represent the Fourier coefficients (straight dotted line) of a diatonic generated scale (polygonal line) for different values of the generating 'fifth', the middle one being the Pythagorean case when $\theta = \log_2(3/2)$.

Many other scales display aligned Fourier coefficients[12], so this does not *characterise* the generated kind. The geometry involved is however reminiscent of Beauguitte's theorem (Theorem 4.7).

---

[12] I presented some alternative cases in the *Helmholtz 'Klang und Ton' Werkshop* in Berlin, 2007. For instance, one can move arbitrarily the first pitch in a generated scale and the Fourier coefficients stay aligned. An even more general parametrisation is a 5-scale with pattern $(0, rt, s - rt, s - t)$.

Fig. 5.4. Fourier coefficients (dots) of a generated seven-note scale (broken line)

5.3 'Diatonicity' of temperaments in archeo-musicology

This third and last notion of DFT, of ordered scales of continuous pitch-classes, provides indicators of 'diatonicity' of a given, non-equal temperament. It is quite difficult to give scientific measurements of the quality of a temperament (or TeT for short), an essentially subjective notion. Among many tries, I will present one that makes use of DFT. It focuses on Bach's well-known enthusiasm for being able to play in the same 'good' or well' (woßl) temperament all major and minor tonalities; quoting the words of the Cantor:

> ... durch alle Tone und Semitonia sowoßl tertiam majorem oder Ut Re Mi anlangend, als auch tertiam minorem oder Re Mi Fa betreffend.

Identifying a tonality with its scale, we can characterise diatonic scales with the following

Theorem 5.6. *Let $\mathscr{S}$ be the set of scales of n notes chosen in some equal tempera-ment with m notes ($m > n$).*
Then the scales in $\mathscr{S}$ with biggest value of $|\mathfrak{S}_{\mathscr{A}}(1)|$ are the Maximally Even Sets.

This is a variant of the definition of ME sets by maximum saliency, cf. Section 4.3. In 12-tone equal temperament, the Maximally Even Scales with seven notes (e.g. the seven-note scales $\mathscr{A}$ with greatest value of $|\mathfrak{S}_{\mathscr{A}}(1)|$) are precisely the 12 major (diatonic) scales. We can see that the difference is substantial by looking at Fig. 5.5, with the DFT profiles of a diatonic scale and another, random scale (notice the small 0th coefficient too, expressing the 'balanced' quality, cf. [67]).

I give the proof of the theorem in the simpler case when the number of notes n is coprime with the cardinality m of the temperament, since for diatonic scales we have $n = 7, m = 12$.

Proof. Since the temperament is equal, we can label the elements of $A \subset \mathbb{R}/\mathbb{Z}$ as $k_j/m, j = 0 \ldots n - 1$ where the k_j are integers, i.e. mA can be seen as a subset of $\mathbb{Z}_m$.

I begin with pointing out that the map $(k, j) \mapsto n \times k - m \times j$ is one-to-one (and onto) from $\mathbb{Z}_m \times \mathbb{Z}_n$ to $\mathbb{Z}_{n \times m}$, where $\mathbb{Z}_p$ stands for the cyclic group with p elements. This morphism (it is well defined, and obviously linear[13]) of $\mathbb{Z}$-modules is injective,

---

[13] It is the canonical isomorphism.

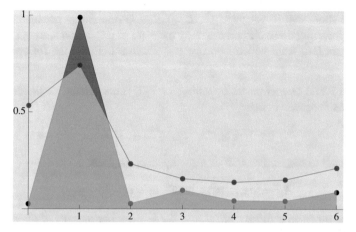

Fig. 5.5. DFT of a major scale vs. another seven-note scale

$$n \times k - m \times j \equiv 0 \pmod{nm} \iff \exists \ell,\ nk = mj + \ell \times mn \iff m \mid k \text{ and } n \mid j$$
$$\iff k \equiv 0 \pmod{m} \text{ and } j \equiv 0 \pmod{n}$$

using Gauss's lemma (m divides nk but is coprime with n, hence divides k, similarly for n), and hence bijective because the cardinalities of domain and codomain are finite and equal.

This enables us to choose n couples $(k_0, 0), (k_1, 1) \ldots (k_{n-1}, n-1)$ in $\mathbb{Z}_m \times \mathbb{Z}_n$ with $nk_j - mj \in \{0, 1, \ldots n-1\} \pmod{m} \times n$, (choosing j first then k_j), hence $\frac{k_j}{m} - \frac{j}{n}$ stays between 0 and $\frac{n-1}{nm} < \frac{1}{m}$.

In order to maximise their sum, the vectors occurring in the computation of $\mathfrak{s}_{\mathscr{A}}(1)$, i.e. the $e^{2i\pi(\frac{k_j}{m} - \frac{j}{n})} = e^{2i\pi \frac{nk_j - mj}{nm}}$, must be as close together as possible: this was proved in the Huddling lemma 4.23.

From the above analysis, the maximum configuration occurs when

$$mA = \{k_0, \ldots k_{n-1}\} \quad \text{with} \quad \{nk_j - mj\} = \{0, 1, \ldots n-1\} \quad \text{(adjoining minimal values)}.$$

Multiplying by $f = n^{-1} \mod m$ yields

$$mA = \{k_j\} \underset{\mod m}{=} \{k_j - f \times m \times j\} = \{0, f, \ldots (n-1)f\},$$

i.e. an arithmetic progression with ratio f, which as we have seen means that A is $\mathrm{ME}_{m,n}$. The most general case is obtained by translation (i.e. a transposition, musically speaking) of this one.

Since DFT is a continuous map, this theorem stays true even for unequal temperaments, which are small perturbations of an equal TeT; though the values fo $|\mathfrak{s}_{\mathscr{A}}(1)|$ may, and will, differ slightly between major scales, these must be the 12 highest values among 7-scales. Let us call *diatonicity* of a diatonic scale A this value $|\mathfrak{s}_{\mathscr{A}}(1)|$.

It is but a short step to consider the *differences in diatonicity* between all 12 and aiming at lowering these differences. Any measure of dispersion among 12 values is suitable[14], in [16] from which I borrow this section, I choose the following definitions:

Definition 5.7. *A temperament, or tuning, or TeT, is an ordered sequence of 12 different notes[15] modulo octave:*

$$0 \leqslant t_0 < t_1 < t_2 < \ldots t_{11} < 1.$$

A major scale in temperament $\mathcal{T}$ is a sequence of the form

$$A_\alpha = (a_0, \ldots a_6) \quad \text{with} \quad a_i = t_{[k_i + \alpha \mod 12]},$$

where α is a constant integer offset and the k_i's are the indexes of the standard C major scale:

$$(k_0, k_1 \ldots k_6) = (0, 2, 4, 5, 7, 9, 11).$$

Example: say $\alpha = 5$, we get the notes a_i with $i = 5, 7, 9, 10, 12 = 0, 14 = 2, 16 = 4$, i.e. F major.

Now we can compute $|\mathscr{F}_{A_\alpha}(1)|$ for all $\alpha = 0 \ldots 11$, i.e. for the 12 major scales in $\mathcal{T}$. For instance, taking for $\mathcal{T}$ the so-called Pythagorean tuning with the 'wolf fifth' between A# and F, we get the following values for all major scales (in semi-tone order starting from C major):

$$0.989, 0.989, 0.986, 0.993, 0.986, 0.991, 0.986, 0.986, 0.991, 0.986, 0.993, 0.986.$$

Notice how close these values are to 1, which illustrates the characterisation of ME sets in Theorem 5.6.

But a most important feature in a given temperament is the *distribution* of these values. In order to visualise this phenomenon more easily, we define

Definition 5.8. *The **Major Scale Similarity** (MSS) of temperament $\mathcal{T}$ is the inverse of the largest discrepancy between diatonicities $|\mathscr{F}_{A_\alpha}(1)|$ for all 12 major scales in $\mathcal{T}$:*

$$MSS(\mathcal{T}) = \frac{1}{\max_\alpha(|\mathscr{F}_{A_\alpha}(1)|) - \min_\alpha(|\mathscr{F}_{A_\alpha}(1)|)}.$$

This quantity is highest when all values of $|\mathscr{F}_{A_\alpha}(1)|$ (for all 12 major scales) are the closest, i.e. when all major scales are almost equally similar to the ideal (theoretical) model of the regular heptagon (Fig. 5.6). For instance for Pythagorean tuning, we get a maximum (resp. minimum) value of 0.993 (resp. 0.986) and hence

$$MSS(\text{Pyth}) = \frac{1}{0.993 - 0.986} = \frac{1}{0.0071} \approx 140.$$

---

[14] And all are equivalent in a topological sense since a vector space with dimension 12 has only equivalent metrics.

[15] The values t_i are computed in practice as intervals (from some arbitrary origin) in cents, divided by 1200 so that one octave = 1. See Section 3.3.

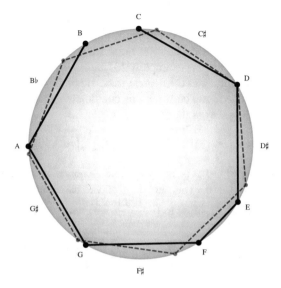

Fig. 5.6. Major scales are best discrete approximations of regular heptagons

For equal TeT of course, all scales are isometric and MMS is infinite. A table of MMS for numerous TeTs can be found in the table section, Fig. 8.36. Since the topic of recovering the TeT used by J.S. Bach is about the most vociferous controversy in music theory[16], I will refrain here from adding any more fuel to the fire, except urging the reader to take into account as many qualities of a given TeT as possible, before selecting the 'best' one – MMS is but one quality among others, albeit more objective than many.

5.4 Fourier vs. voice leading distances

We have just recalled the importance of the closeness to some regular division of the chromatic circle.

But the following statement, obtained by D. Tymoczko [88] through comprehensive computations, is surprisingly precise:

Proposition 5.9. *The magnitude of a chord's d^{th} Fourier component is closely correlated to the size of the* minimal voice leading[17] *from the chord to the closest subset of any perfectly even n-note chord.*

NB: here we are back to the alternative definition of magnitude of DFT introduced by Tymoczko on n-note orbifolds (see Section 5.1), i.e. the pitch-classes of the 'perfectly even n-note chord' need not be integers.

---

[16] See www.larips.com for instance.

[17] For Euclidean distance, see below.

Of course it would be excessive to interpret this as a subordination of Fourier computation to voice leadings: to begin with, finding 'the closest subset of a perfectly even chord' requires a calculation remarkably akin to that of the phase of Fourier coefficients! To quote his examples, this means for instance that the third coefficient is close in magnitude to the nearest translate of $\{0\}, \{0,4\}$ or $\{0,4,8\}$. The experimental equation he found between this coefficient's magnitude and the length (in Euclidean quotient space) of the voice leading is

$$VL \approx -0.64 \times |\mathscr{F}_A(3)| + 2.12 \iff |\mathscr{F}_A(3)| \approx 3.39 - 1.57 \times VL.$$

(The equation given in [88] had to be rewritten since Tymoczko uses a different convention for the Fourier transform. This does not alter the quality of the correlation.)

For instance for $\{0,4,7\}$, close to $(-\frac{1}{3}, \frac{11}{3}, \frac{23}{3})$ with

$$VL = \left\| (0,4,7) - (-\frac{1}{3}, \frac{11}{3}, \frac{23}{3}) \right\| = \sqrt{\frac{1}{9} + \frac{1}{9} + \frac{4}{9}} = \sqrt{2/3} \approx 0.816,$$

we get the approximate value $|\mathscr{F}_A(3)| \approx 2.10$ instead of the exact value $\sqrt{5} \approx 2.23$.

This approximation (and similar ones for other coefficients) attains very high correlation coefficients, mostly in the $[-0.99, -0.97]$ range (depending on the cardinality of the pc-set and the index of the coefficient). It is indeed intuitive that the closest we come to a maximum, the greatest the value of the Fourier coefficient, which does go some way into explaining a correlation. However, correlations are sometimes misleading[18] and indeed, there are several caveats in this:

- Near a maximum, a map moves horizontally not obliquely (the derivative is 0). Specifically, one expects to reach exactly value d (3 in the above formula instead of 3.39) when VL= 0, and keep a nearly horizontal slope close to the maximum.
- Moving away from a maximum implies that any point will be *below the maximum*, not that the map is *globally decreasing* – a common fallacy.
- Why restrict the statistic to genuine (discrete) pc-sets, when one has chosen to work with non-integer pcs?

In [89] Tymoczko states that 'it would be possible (...) to calculate this correlation analytically.' I proceeded to do so, but the result is not the same as his.[19]

Let $A = (a,b,c)$ be for simplicity's sake a 3-subset of $\mathbb{Z}_{12}$, and assume furthermore that B, 'the closest subset of any perfectly even chord', has the form $(x-4, x, x+4)$, i.e. informally $a \approx x-4, b \approx x, c \approx x+4$. The calculation is similar when B has type $(x, x, x+4)$ and other cases, and indeed for any d-subset in any n-TeT, see the general formula in Proposition 5.10 below.[20]

---

[18] According to Mark Twain, famous Victorian British PM Benjamin Disraeli once declared:'There are three kinds of lies: lies, damned lies and statistics.'

[19] This was first presented in [12].

[20] Computationally the difficult part is to identify what *type* of subset of a regular polygon is closest to A. This can be done in polynomial time, checking and comparing possible types,

To begin with, the closest such B occurs when $x = \frac{a+b+c}{3}$ (assuming $0 \leqslant a \leqslant b \leqslant c < n$ again for clarity). This follows from the study of the square of distance AB, the voice leading distance:

$$VL^2 = AB^2 = (a-x+4)^2 + (b-x)^2 + (c-x-4)^2 = f(x) \qquad \frac{\mathrm{d}f(x)}{\mathrm{d}x} = 2(a+b+c-3x).$$

Remember that when the minimum distance is reached, the sum of the angular differences between A and B, $(a-x+4) + (b-x) + (c-x-4)$, is nil.

In Fig. 5.7 we see that the perfect division in 3 closest to C major is $(-1/3, 11/3, 23/3)$.

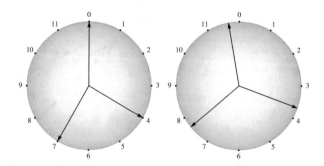

Fig. 5.7. Approximating $(0,4,7)$

From now on we assume this value for x. Let us compute $\mathscr{F}_A(3)$ with the idea in mind that $a - x + 4$ and similar quantities are 'small'.[21]

$$\mathscr{F}_A(3) = e^{-2i\pi 3a/12} + e^{-2i\pi 3b/12} + e^{-2i\pi 3c/12} = e^{-2i\pi a/4} + e^{-2i\pi b/4} + e^{-2i\pi c/4}$$

$$= e^{-2i\pi x/4}\left(e^{-2i\pi(a-x+4)/4} + e^{-2i\pi(b-x)/4} + e^{-2i\pi(c-x-4)/4}\right)$$

$$= e^{-i\pi x/2}\left(e^{-i\varphi} + e^{-i\psi} + e^{+i(\varphi+\psi)}\right),$$

setting $\varphi = \dfrac{\pi}{2}(a-x+4), \psi = \dfrac{\pi}{2}(b-x), -\varphi - \psi = \dfrac{\pi}{2}(c-x-4)$ according to the definition of x.

But from the power expansion of $e^{it} = 1 + it - \dfrac{t^2}{2} + \ldots$, one gets

$$e^{-i\varphi} + e^{-i\psi} + e^{+i(\varphi+\psi)} \approx 3 - \frac{1}{2}\left(\varphi^2 + \psi^2 + (\varphi+\psi)^2\right) \in \mathbb{R}.$$

Hence, since $|e^{-i\pi x/2}| = 1$,

---

e.g. $(x-4, x, x+4), (x, x, x\pm 4)$ and (x, x, x) for 3-subsets. In practice, one also has to keep in mind that the computations are modulo n, e.g. a number such as 11.57 is probably best construed as $-0.43 \mod 12$.

[21] This enables us to resolve the ambiguities, in particular that x is only defined modulo $n/3$: there is one ordering of B that is closest to A, and it is the one we are interested in.

$$|\mathscr{F}_A(3)| \approx 3 - \frac{1}{2}(\varphi^2 + \psi^2 + (\varphi + \psi)^2) = 3 - \frac{\pi^2}{8} VL^2.$$

Of course this formula is again an approximation, quite good for $VL < 1$ (but meaningless when $VL^2 > 24/\pi^2$)[22]; it suggests looking for a correlation with VL^2 instead of VL, which would provide as good a fit as [88], or better.

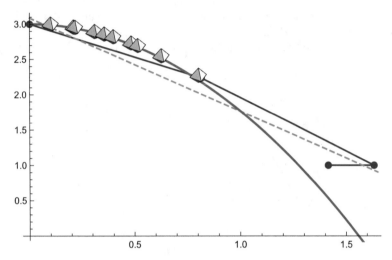

Fig. 5.8. $|\mathscr{F}_A(3)|$ is quadratic in VL, not linear.

The calculation above can obviously be carried to divisions in d parts instead of 3, to $n \neq 12$, and to all cases of sub(multi-)sets of an even division (this was done for producing Fig. 5.8). The general formula is the following:

Proposition 5.10. *For any d-subset in a chromatic universe with n pcs, if VL is the Euclidean voice-leading distance to the nearest sub(multi)set of a perfect division of the continuous circle in d, one has*

$$|\mathscr{F}_A(d)| \approx d - \frac{2d^2 \pi^2}{n^2} VL^2.$$

Of course this equals d when the pc-(multi)set *is* a perfect division of the circle, which is the expected value, and the approximation is better the closer one gets to such an evenly divided subset.

The general proof is very similar and is left as an exercise.

This analytic expression, like any approximation formula, can go awry when one gets far from even pc-sets ('rogue' chords).[23] This is less apparent with Tymoczko's

---

[22] NB: pushing the Taylor expansion further would show that the next term is $\frac{\pi^4}{172}(\varphi^4 + \psi^4 + (\varphi + \psi)^4)$ – the third-order term also vanishes. This is specific to 3-subsets.

[23] The neglected terms may be substantial for largish values of d: for instance for the diatonic scale the exact value of $|\mathscr{F}_A(7)|$ is $2 + \sqrt{3} \approx 3.73$ whereas the approximate formula yields

linear regression, but artificially so since it is computed by giving equal weight to pc-sets that do not approximate well an even distribution. On the other hand, introducing random pc-sets with non-integer values (whose presence is the whole point of the orbifolds models!) vindicates the second-order formula. Such sets have been randomly added to Fig. 5.8 and appear as tiny octahedra; they are clearly much better fitted by the second-order formula than by the linear interpolation.

The aim here was to reach a non-heuristic, more precise understanding of why and how the Fourier coefficient should decrease when the voice leading to the closest even subdivision augments. Though I would not hierarchise voice leadings vs. DFT as Tymoczko does, we both proved albeit in different ways that it is the *Euclidean* metric which best correlates the two notions.

5.5 Playing in Fourier space

So far we have seen the DFT, and the less discrete continuous FT, used for analysis of musical structures such as pc-sets in musical pieces. It is time to take a look at its creative power and virtue. For one thing, the neat separation of recognisable musical qualities into different Fourier coefficients means that modification on the fly of one coefficient will trigger the musical quality that it embodies, without having to look at the separate dimensions of, say, the quantities of each pitch-class involved. Besides, enhancing the quality attached to one Fourier coefficient is done efficiently, without wasting energy on extraneous dimensions. In terms of complexity, changing for instance the diatonicity of a pc-set involves changing all or almost all 12 pitch-classes truth values, whereas an increase of the sole fifth Fourier coefficient is sufficient for the same effect.

Several experiments have been conducted in moving directly and purposefully in Fourier space. The last that I will discuss addresses the psycho-acoustic perception of saliency as defined by the distribution of Fourier profiles.

5.5.1 Fourier scratching

In the Yale conference for the Society of Mathematics and Computation in Music (2009) two practical, creative applications of Fourier spaces were presented in the final panel. The 'Fourier scratching' created by Thomas Noll and Martin Carlé[24] is best seen in action before explained and the reader is strongly encouraged to have a look at https://youtu.be/6HipqANRXPY before carrying on with reading.

Fig. 5.9 is a pale substitute for attending an actual performance of Fourier scratching: the DJ uses controllers (actually a game pad) to modify interactively

---

3.16. Most of the error is in this fourth-order term, here ≈ 0.61. In general, the approximation is acceptable when $d < n/3$ or so, and it is advisable to compute the Fourier coefficient of *the complement* of any pc-set whose cardinality exceeds $n/2$: for the complement of the diatonic scale, the approximation formula yields 3.63, a much better result.

[24] It expanded in a spectacular way the previous, tentative experiment in [69].

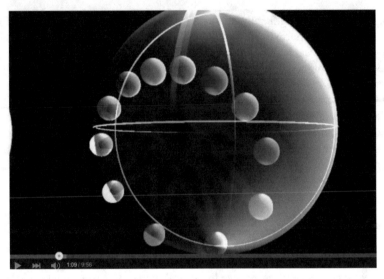

Fig. 5.9. A snapshot of Fourier scratching on a 12 note-rhythm

parameters in Fourier spaces, which change in real time the production of a periodic rhythm while these parameters are projected on screen. The actual implementation is explained in great detail in [70] and in this section I will only provide a brief overview.

The DJ starts from a predefined 'rhythmic loop' which is simply a cyclic loop of musical events, each parameterised by two dimensions: $s_0, s_1, \ldots s_n = s_0, s_{n+1} = s_1, \ldots$ where the s_i are complex numbers. $|s_i|$ is naturally interpreted as the loudness of the sound event and $\arg(s_i)$, an angle, can be used in a variety of paradigms (see below how it can be used as a choice of musical scale) but in the Yale demonstration the most impressive effect was FM, changing the colour of the sound.

DFT is applied to the *sequence* $(s_0, \ldots s_{n-1}) \in \mathbb{C}^n$ (according to Noll's definition of DFT in 5.2 above), providing a cycle of Fourier coefficients $(a_0, \ldots a_{n-1}) \in \mathbb{C}^n$ where $a_k = \sum_{j=0}^{n-1} s_j e^{-2ijk\pi/n}$. What we see in Fig. 5.9 is a stereographic projection of these n coefficients[25], and the DJ acts with controllers *on the a_i* which by inverse DFT modifies the rhythmic loop in real time, in a way analogous to the 'scratching' of a more conventional DJ, accelerating or slowing the reading of a LP. Each sphere is highlighted in turn, though none is associated with a particular sound event – quite the contrary: all take into account, and hence modify, *the whole of the rhythm* (for instance, coefficient a_0 is the bare sum of all the s_i and hence its proximity to south pole is a measure of the 'balance' of the parameters of the original rhythmic loop. It takes little experimenting to grasp the meaning of these coefficients).

---

[25] Roughly speaking, a complex number on this spheric representation is the larger when it is closer to the north pole, and its phase is given by its longitude.

I can do no better than reproduce[26] an example taken in [70]: in Fig. 5.10 we can see a very simple Fourier configuration: all coefficients are nil except one, because the cyclic rhythm is a regular polygon centered on the origin (e.g. a metronome). Scratching a single Fourier coefficient provides a rich, complex modification of the rhythm: on the third picture we can figure the binary beat induced by every odd event happening farther from the origin, meaning greater loudness.

Pictures on paper are but a pale substitute for the real thing. Playing with the controls is a memorable experience that enables one to understand quickly the meaning of each Fourier coefficient. Also this demonstration proved beyond any doubt that *thinking* in Fourier space is a good way to address in one go (playing on just one coefficient) complex but recognisable musical features.

Fourier scratching was also used in the much more complex and theoretical context of 'scale labyrinths' by the same authors and a few others [68], as a good and practical way of experimenting with diverse well-formed scales in specific tunings. Again, sequences of notes are played and repeated in a given scale, and the performer or DJ or 'scratcher' directly manipulates the spectra (DFT) of these sequences, visualised on a sphere. I single out the following quotation in [68]:

> While we cannot offer empirical evidence yet that this particular technique is musically more effective than other alternatives, it is useful to observe that the partials [...] correspond to musically elementary patterns. [...] Early experiments with this system give the impression that play states which are closely related in their Fourier coefficients are sensibly related by the musical ear.

This is yet another vindication of the psycho-acoustic pertinence of DFT, see 5.5.3 below. It is not surprising that Pierre Beauguitte also devised a pedagogical Fourier scratching of rhythms, for an exhibition at the *Palais de la Découverte* in Paris (along with the MCM 2011 conference), see [23] p. 11.

5.5.2 Creation in Fourier space

Another practical realisation of moving in Fourier space presented at MCM 2009 in Yale was the ballet *Dancing the Violent Body of Sound* by Diyah Larasati and Dag Yngvesson, a project instigated and energetically led to completion by Guerino Mazzola at the University of Minnesota. Images of the ballet and of its rehearsals and more can be found at https://vimeo.com/9565561 and will, as usual, give plenty more than mere words can convey. Quoting the booklet:

> [Everything starts from] the seminal case of Fourier decomposition of a sound. It is embodied (the word is apt) by the movements of dancers, each impersonating a $c_k e^{2i\pi kt/n}$. This rotating technique is inspired from East Java traditional dance practice, which Rachmi Diyah Larasati had to teach to the

---

[26] Thanks to Thomas Noll for permission to reproduce.

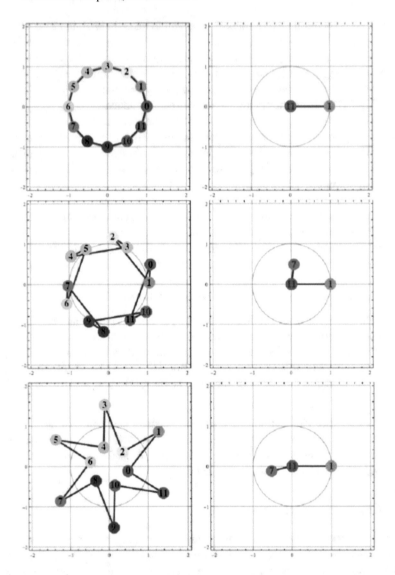

Fig. 5.10. Modifying a single coefficient

dancers to begin with. The jumping and walking movements, and even the different rotation techniques, are obviously in relation with the common Indonesian martial art practice, as we will see quite well in some parts. Captors on the dancers' wrists allow to turn these movements into music via a computer system: on each wrist, a flex sensor controls the sound volume of a partial, an accelerometer controls its frequency. Hence the dancers distort the harmonics of a cello recording by Schuyler Tsuda. During rehearsals, the dancers learned to use the sensors in order to fine-tune their movement's tempo.[27] Variety in the dance is obtained through the violent interventions of 'police dancers' — the two guys in saffron tunics, who execute a kind of 'kata' (combination of martial moves). This idea is dear to the choreograph, Diyah Larasati, who also appears as a free agent during the dance.[28]

5.5.3 Psycho-acoustic experimentation

Last but not least, Pierre Beauguitte's master's dissertation [23] addresses in its final section a crucial issue, the perception of 'Fourier qualities'. Unfortunately this work is still embryonic – there is in it the substance for several PhDs – but still it puts forth convincing evidence that the saliency (as defined by Quinn), at least, is a recognisable quality for the human brain.

Initially I had envisioned (together with Moreno Andreatta and Carlos Agon in IRCAM) a listening test for the perception of periodic rhythms or scales with some high Fourier coefficient; for instance, modulo 12 both $(0, 2, 4, 7, 9)$ and $(0, 1, 3, 4, 6, 7, 9, 10)$ can be considered salient (being among Quinn's prototypes) but $(0, 1, 4, 6)$ is not (being a FLID). Informal experimentation suggested that this is indeed a recognisable characteristic (notice that it overlaps periodicity but remains quite distinguishable from it, cf. the first example given).

After discussion with some psycho-acousticians, specialists of testing and its biases, Beauguitte preferred to test the perception of similarity (of DFT profiles) of periodic rhythms by a panel of listeners. His first conceptual argument is a very strong one that we have already encountered (we will return to it in Chapter 6 where pc-sets are compared by the phases of some Fourier coefficients): granted that the rhythms share the same period n, it is possible to compare the DFTs of two rhythms even when they have different *cardinalities*, since the DFTs are elements of the space $\mathbb{C}^n$. Beauguitte cites rhythms $A = (0, 4, 8)$ and $B = (0, 1, 4, 5, 8, 9)$ in $\mathbb{Z}_{12}$ which are rather difficult to compare[29] in traditional frameworks (see [87] for a comprehensive survey) though the Fourier profiles are readily computed and their distance measured, see Fig. 5.11.

Beauguitte computes the L^1 mutual distances of the *saliencies* in 20 different groups of five rhythms:

---

[27] The most obvious correlation is between the 'fundamental' dancers (in red dress) and the loudness of the bass sounds.

[28] Diyah Larasati is particularly interested in the relationship between violence, dance and embodiment in Indonesia. The cello composition with electronic distortion via the Max software is the work of Schuyler Tsuda. Mathematician and musician William Messing collaborated with Mazzola on the mathematical aspects of the project. The professional recording on video is due to Dag Yngvesson. Toni Pierce-Sands was responsible for rehearsals.

[29] Even more so with a slight modification of this example, say $B = (0, 1, 4, 5, 9)$. In this example the distance is equal to 5.485, a comparatively short value that corresponds to their intuitive proximity.

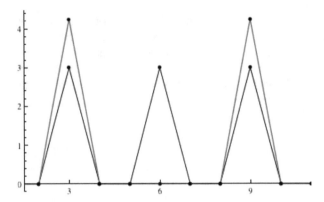

Fig. 5.11. Saliencies of two similar rhythms with different cardinalities

$$\|A - B\|_1 = \sum_{k=1}^{n} \left| |\mathcal{F}_A(k)| - |\mathcal{F}_B(k)| \right|,$$

where the 0th coefficient (cardinality) is purposefully omitted, and the distance is 0 when A, B are homometric. For each group, 21 listeners were asked to measure (moving a cursor) the degree of similarity between a reference rhythm and the four others.

The correlation between the evaluations of listeners and the homometry distance is rather good (coefficient 0.78), though Beauguitte noticed a large dispersion and wondered whether listeners are sensitive to T/I equivalence rather than homometry. A second test with similar protocol demonstrated that homometry is perceptible even when there is no T/I relationship, though the correlation is weaker (55% when random answers would yield 33%) and seems stronger for complementary rhythms (Babbitt's theorem).

These tests should be of course furthered and enhanced[30] but so far support the evidence that the size of Fourier coefficients (saliency) is perceived by the human brain. In itself, this is indicative that the topic of the present book is more than an abstract theorisation, or another pretty tool; DFT begins to bridge the gap between intellectual concepts and perception.[31]

Exercises

Exercise 5.11. (Orbifolds) Find a *fundamental domain* for triads under transposition and inversion: a set that contains one and only one point for each class of triads mod-

---

[30] Especially in the form initially proposed.

[31] It has been established that a living brain is mandatory for listening to music, the mechanism of the cochlea alone being insufficient. It is also well known that some regions in the brain are quite capable of Fourier analysis, though so far this was essentially studied, after Helmholtz, for the perception of pitch.

ulo the action of the continuous group T/I (e.g., there is one point for all major/minor triads). Hint: a triangle should suffice.

Exercise 5.12. (Simultaneous rational approximation) Find a common approximation for $(\sqrt{2}, \pi)$ with $N = 100$ in Lemma 5.2.

Exercise 5.13. Check the computation of Fourier coefficients for Callender's pc-sets P, Q, R, S.

Exercise 5.14. Check that modifying the first pc of a generated scale still keeps the Fourier coefficients aligned (using Noll's DFT of an ordered sequence).

Exercise 5.15. Pick a random trichord A (with non-integer values). Find the closest equilateral set or sub(multi-)set of one, measure the distance between the two and compare with $|\mathscr{F}_A(3)|$ according both to Proposition 5.10 and Tymoczko's linear formula.

Exercise 5.16. Derive the general formula in Proposition 5.10.

6

Phases of Fourier Coefficients

Summary. We have explored in great depth one dimension of Fourier coefficients, their magnitude. This has proved a worthwhile journey, with incontrovertible musical meaning; it allows the painting of nice pictures of scales/chords landscapes, though with the major and embarrassing restriction that scales must share their cardinality in pictures such as Fig. 5.3; also the phase component had to be discarded because it did not make sense in most orbifold universes. It is now time to get back to genuine, ordinary pc-sets and look at the entirety of Fourier coefficients, taking into account not only their magnitudes but also their directions (or 'phases'). This has been tackled in different ways, the first comprehensive try being Justin Hoffman's in [50], developing upon a remark of Joseph Strauss. However I will devote the bulk of this chapter to the study of phases *per se*, since the magnitude has been previously covered extensively.

I will only provide a few chosen musical examples, the purpose of this book being rather a clean and comprehensive exposition of the theoretical background necessary for such endeavours. The torus of phases was introduced in [15], but I refer the reader to Yust's pioneering work for many far more convincing analyses, cf. [96, 97, 98, 99, 100].

6.1 Moving one Fourier coefficient

Depending on the index and the cardinality of sets, the landscape of all values of one Fourier coefficient takes on strange shapes. See the haunting picture for a_5 on all 3-sets on the left of Fig. 6.1 with major/minor triads coloured in blue. The picture for a_3, right, is much more barren. Both are inspired from [50] but were redrawn for the present opus.

In [50], Hoffman introduced an intriguing description of those spaces, focusing on parsimonious moves: whenever C is moved to C♯ (say) in a pc-set, the Fourier coefficient with index k will be changed by the same quantity, $e^{-2ik\pi/12} - e^0$, which provides the picture with a partial lattice structure – partial because only those pc-sets containing C can be translated by this vector.

Additional symmetries occur because all similar semitone moves (say D to D♯) are deduced from the one above by a rotation:

$$(D\sharp - D)_k = e^{-2ik3\pi/12} - e^{-2ik2\pi/12} = e^{-2ik2\pi/12} \times (e^{-2ik\pi/12} - e^0).$$

© Springer International Publishing Switzerland 2016
E. Amiot, *Music Through Fourier Space*, Computational Music Science,
DOI 10.1007/978-3-319-45581-5_6

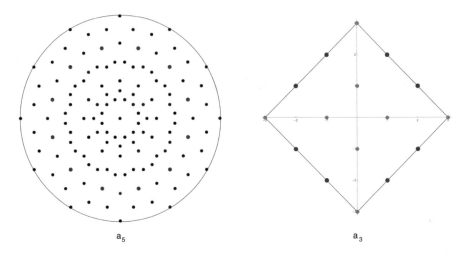

Fig. 6.1. a_5 and a_3 coefficients for all 3-sets in $\mathbb{Z}_{12}$

In general,

Proposition 6.1. *Vectors carrying a pc-set to another one obtained by a fixed parsi-monious move (meaning one pc and one only is moved by some minimal amount, a semitone or a tone for most theorists) take their values on a regular polygon.*

Proof. Without loss of generality, say we change a pc-set by moving one of its pcs by a single semitone. Algebraically, in the space of distributions, we add (to characteristic functions) the new value of the pc and subtract the original one, i.e. we add (say) D♯ and remove D. By linearity of the DFT, this changes the Fourier coefficient by the quantity computed above, taking values in rotations of C♯ - C (the number of different rotations depends on the index of the Fourier transform and the number of semitones of the voice-leading).

Indeed, in complex numbers the values of these differences of coefficients a_k between pc-sets are $\pm 2 \sin \frac{k\pi}{n} e^{i(2p-1)k\pi/n}$ where only p varies. This explains these almost complete regular polygons on Hoffman's pictures, cycling from one pc-set to the next by the same movement (while this movement is permissible). Moreover, since the change in pc-set space is small, by continuity of DFT the change in Fourier space is also small and more often than not it is a minimal move, linking neighbours. That good voice-leadings be recognisable as such in Fourier space is important, but these spaces also enable the visualisation of broader concepts of proximity as we will see below.

Finally, as can be seen from the paucity of points in the right half of Fig. 6.1, the computation of a_k is not injective: several pc-sets may project onto the same point, for instance C major and D minor triads share the same coefficient a_5.[1]

This explains the difference in complexity: semitone displacements of one pc, for instance, reduce to only four possible translations when $k = 3$; whereas there are 12 different cases for $k = 1$ or $k = 5$. This would produce complete regular dodecagons on the picture, if multisets were allowed, as in Fig. 6.2. For instance, between C major (D minor) and G major (A minor) triads, one finds three-note pc-multisets such as FAA or $\{5,9,9\}$ (which coincides with $\{7,11,11\}, \{7,7,11\}$ and $\{5,5,9\}$) together with the more mundane $\{0,2,4\}$ or CDE. There are common notes and symmetries between these pc-sets, which will be better explained in 6.3.1.

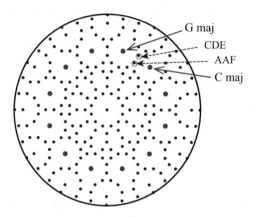

Fig. 6.2. a_5 coefficient for all 3-**multi**sets in $\mathbb{Z}_{12}$. Major/minor triads are the large blue dots.

Quinn's approach is focused on the distance between the points on these pictures and the origin, purposely excluding any consideration of angular position. After [50], [15] and [96] have begun to explore this remaining dimension. Whereas the *magnitude* of Fourier coefficients has taught us much about *shape* (being rigorously equivalent to the internal intervallic distribution) and the kinship of a given pc-set with some generated scale – its saliency for a specific index and cardinality – we will see that the *phases* of Fourier coefficients tell much about *harmony* and modulations.

6.2 Focusing on phases

The difficulty of working in $\mathscr{F}_A(k)$ space with fixed k is threefold at least: its complexity, the lack of injectivity (too many pc-sets project on the same coefficient, es-

---

[1] This may appear as a drawback; however, see below in Section 6.2.2 how two coordinates allow us to discern between triads, and consider that in the diatonic universe of C major, generated by the interval 5, all pcs of the D minor triad belong to the scale of C major.

pecially when $k \mid n$) and the fact that it does not clearly reflect the cyclicity of the original spaces, $\mathbb{Z}_n$ and its subsets.

Besides, the magnitude of Fourier coefficients has demonstrated its importance and validity, following Quinn's seminal work about 'chord quality' and saliency. It looked like a good idea therefore to explore the meaning of the other component: phase. In the DFT of a pc-set $A \in \mathbb{Z}_{12}$ there are only six significant phases, since the phase of $\mathscr{F}_A(0)$ is always 0 and the phases of $\mathscr{F}_A(7), \mathscr{F}_A(8) \dots$ are opposite to the phases of $\mathscr{F}_A(5), \mathscr{F}_A(4) \cdots$. Even the phase of coefficient $\mathscr{F}_A(6)$ is of limited interest since this coefficient is an integer, the difference between the number of even and odd elements in A.

In the seminal [15], several choices of two indexes were tried, argued, and compared; it was advocated that consideration of $\mathscr{F}_A(3), \mathscr{F}_A(5)$ is a good choice at least for diatonic music, since it enables differentiation between all 24 major/minor triads[2], and as we will see below, it embeds the usual Riemannian (dual) Tonnetz in a space representing all (or most) pc-sets with any cardinality. [96] proved that the corresponding space is an excellent one for the description of early romantic music, the ambiguities of the model (like diminished seventh having no definite coordinate) even mirroring harmonic perceptions. [98] extends the analysis of phases (possibly with different coefficient indexes) to segmentation of passages of 20th century music, see Section 6.2.2. The reader will not meet any difficulty in adapting the following to other torii.

6.2.1 Defining the torus of phases

Phase intervenes naturally already in the computation of magnitudes of Fourier coefficient: as we have seen in some previous examples, the magnitude of $\mathscr{F}_{A+A'}$ is not necessarily the sum of the magnitudes of $\mathscr{F}_A$ and $\mathscr{F}_{A'}$. Depending on the direction of the coefficients, they can even cancel each other out (cf. the Guidonian hexachord Fig. 8.26 where $a_6 = 0$ because the two whole-tone chunks cancel each other out, or a_3 in the decomposition of the first trope of Webern's *Variationen* in Fig. 4.21, or the disintegration of the diatonic character when reuniting the three instruments in Stravinsky's *Three Pieces* discussed again below, Fig. 6.9). Yust spelled out the explicit formula ([98]):

Proposition 6.2. *Let $B = A \cup A' \cup \dots$ meaning $1_B = 1_A + 1_{A'} + \dots$ Then*

$$|b_k| = |a_k| \cos(\varphi_{b_k} - \varphi_{a_k}) + |a'_k| \cos(\varphi_{b_k} - \varphi_{a'_k}) + \dots$$

where $a_k = |a_k| e^{i\varphi_{a_k}}$ (resp. b_k, etc.) denotes the kth Fourier coefficient of A with phase φ_{a_k} (resp. phase φ_{b_k}).

This formula (obtained by projection on the direction of b_k in the complex plane, cf. the proof of Huddling Lemma 4.23) pinpoints that b_k increases when a_k points in the

---

[2] Actually, for any homometric class (e.g. major/minor triads), only the phase of the Fourier coefficients changes; and focusing on phases is the best way (fewer coordinates) to differentiate between homometric elements.

same direction – when φ_{a_k} is close to φ_{b_k} – does not change when a_k is at a right angle to b_k and diminishes when a_k points opposite to the sum. The importance of phases in combinations of pc-sets being paramount, it is only natural to have a look at phases *per se*. In order both to allow pc-sets with any cardinality and to be able to distinguish between them, it proves best to consider at least two dimensions:

Definition 6.3. *The* torus coordinates *of* $A \subset \mathbb{Z}_{12}$ *are the two angles* $\varphi_3 = \arg(\mathscr{F}_A(3))$ *and* $\varphi_5 = \arg(\mathscr{F}_A(5))$. *The* torus *of phases is the space* $\left(\mathbb{R}/(2\pi\mathbb{Z})\right)^2$ *of all such pairs of angular coordinates.*[3]

Example 6.4. Consider a simple pc-set: $A = \{0,2,4\}$, i.e. CDE. Then

$$a_3 = e^0 + e^{-2\mathrm{i}6\pi/12} + e^{-4\mathrm{i}6\pi/12} = 1 - 1 + 1 = 1,$$
$$a_5 = e^0 + e^{-2\mathrm{i}10\pi/12} + e^{-4\mathrm{i}10\pi/12} = 1 + e^{\mathrm{i}\pi/3} + e^{2\mathrm{i}\pi/3} = 1 + \mathrm{i}\sqrt{3} = 2e^{\mathrm{i}\pi/3}$$

hence $(\varphi_3, \varphi_5) = (0, \pi/3)$.

This phase space is topologically a torus, though I have been convinced that Yust's planar representations are easier to understand than the traditional doughnut, provided one remembers to identify opposite sides of the picture (compare Fig. 6.3 and 6.4 below). Since this visualisation is customary nowadays in neo-Riemannian analysis, we will usually follow this trend in this chapter.

It is important to acknowledge that some collections do not admit torus coordinates: for instance, a tritone T has $\mathscr{F}_T(2k+1) = 0$ for all k, so that both angular coordinates are undefined; any union of tritones (such as a diminished seventh) shares the same indignity. *Idem* for a whole-tone scale. An augmented triad fares a little better, since φ_3 exists, but φ_5 does not (ergo such pc-sets can be construed if necessary as vertical lines in Fig. 6.3).

That the torus of phases is concerned with harmony (while amplitude of Fourier coefficients is all about shape) is apparent from the following corollary of theorem 1.16:

Lemma 6.5. *Transposition of a pc-set by t semitones rotates its kth Fourier coefficient a_k by a $-2kt\pi/n$ angle, i.e.* $\varphi_k \mapsto \varphi_k - 2kt\pi/n$.

Any inversion of a pc-set similarly rotates the conjugates of the Fourier coefficients.

Proof.

$$\widehat{\mathscr{F}_{A+t}}(k) = \sum_{m\in(A+t)} e^{-2\mathrm{i}\pi km/n} = \sum_{m-t\in A} e^{-2\mathrm{i}\pi k(m-t)/n} e^{-2\mathrm{i}\pi kt/n} = e^{-2\mathrm{i}\pi kt/n} \times \mathscr{F}_A(k).$$

Similarly for inversion, one gets for instance (remember $-A$ means the inverse of A around pc 0):

---

[3] These angles are defined modulo 2π, as is customary in planar geometry; however, Yust may well be right in changing the modulo to 12 by a simple multiplication $\varphi \mapsto \frac{6}{\pi}\varphi$, since this enables recognition of these angles in reference to single pcs.

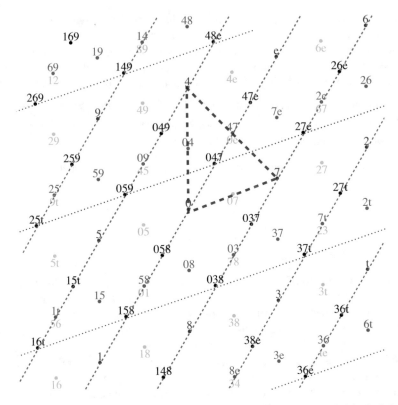

Fig. 6.3. Major/minor triads, dyads (thirds in red), single pcs alias diminished triads

$$\widehat{\mathscr{F}}_{-A}(k) = \sum_{-m \in A} e^{-2i\pi k(-m)/n} = \overline{\mathscr{F}_A(k)}.$$

For instance, transposition by fifth of a diatonic collection (or major/minor triad) changes φ_5 by the smallest increment possible, $\pi/6$. One can actually draw a circle with the values of φ_5 (which are the directions of the points in Hoffman's diagram 6.2), and use it to correlate any pc-set with a diatonic scale (or, say, major triads), cf. the Stravinsky example in Fig. 6.9.

Since any transposition means a shift in both phases according to Lemma 6.5, all transpositions are global operations and appear as translations on the diagram. For instance, transposition by a major third appears as a vertical translation by $2\pi/3$. Because the plane is not infinite but a quotient space, where opposite sides of the picture are glued together, the orbit of a translation usually appears as several disjunct lines which are actually the same one (see Figs. 6.4 and 6.12).

Single pcs (or major triads, or diatonic collections, etc.) could actually be distinguished by φ_5 alone since when $A = \{k\}$, $\varphi_5 = 10k\pi/12$ takes on 12 different values modulo 2π (5 being coprime with 12) (whereas $\varphi_3 = k\pi/2$ cycles through four values only; similarly for the orbit of any given pc-set under translation).

Proposition 6.6. *The 12 pcs* $(k\pi/2, 5k\pi/6)$ *are aligned on the* chromatic line $5\varphi_3 = 3\varphi_5$ *with slope* $5/3$.

Let us stress again that this apparently broken line is actually connected and circular, since it is drawn on a torus, cf. Fig. 6.4. It appeared in blue in Fig. 6.3.

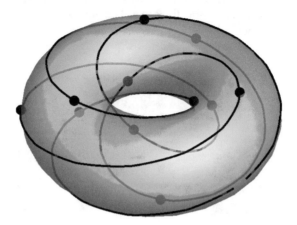

Fig. 6.4. The 'chromatic straight line' of single pcs

Remember also (by the Tritone Lemma 3.9) that these single notes coincide with diminished triads: 6 is also the position of $\{3, 6, 9\}$.[4]

More generally, coefficients and *a fortiori* their phases may be the same even for pc-sets with different cardinalities: from the Tritone Lemma 3.9 proved above, *when they differ by adding or removing tritones, two pc-sets will share the same pair of torus coordinates.* This happens for instance for a diatonic collection and the associated pentatonic (C major and CDEGA) or a dominant seventh and its fifth (CEGB♭ and CG).[5] Hence the torus of phases is not a faithful representation of *all* pc-sets; but

1. Like Hoffman's spaces, and unlike orbifolds or voice-leading spaces which stick to one cardinality for subsets, it allows us to view and compare sets and even multisets with any cardinalities in the same space (with fixed, small dimension).
2. The singular subsets (i.e. without definite coordinates) are few and far between.[6]

---

[4] In the same line, I appreciate the identification of the incomplete dominant seventh with the fifth degree: CDF♯ shares the same a_3, a_5 as D alone.

[5] Remember however that this condition is sufficient but not necessary. The phase coordinates φ_3, φ_5 of a diatonic CDEFGAB, pentatonic CDEGA and major third on the tonic CE are equal, and φ_5 is still the same for the second degree D alone (though the Fourier *coefficients* are different, because their magnitudes differ). Yust also lists harmonic minor scales and their tonic triads, and a few other cases, cf. infra.

[6] As shown in the last example in [15], if it is desirable to visualise diminished sevenths, then consideration of φ_4 instead of φ_3 will do the trick. This may also be considered if 4-chords

3. Lacking injectivity is an asset, not a drawback – I am happy to confuse CDE-FGAB with CDEGA[7], or a minor scale and its triad, or sharing the same φ_4 among minor-third transpositions of a seventh chord. See below in Lemma 6.7 the meaning of the confusion between minor thirds and seconds.

4. *The disposition of triads is topologically the same as the Riemannian Tonnetz* (cf. Figs. 6.3 and 6.5): the neighbours of a given triad are its images by the famous PLR local transformations, e.g. C major is surrounded by E minor, A minor and C minor.

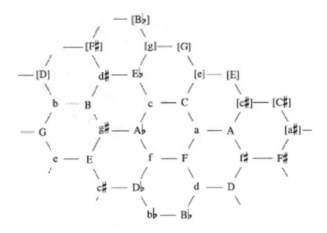

Fig. 6.5. The dual tonnetz of triads

5. There is some psycho-acoustic validation of this topological space: quoting [96], The Tonnetz-like arrangement of triads in [the torus of phases] is approximately equivalent (up to some scaling of parameters) to the experimentally derived space of key relationships presented in [58].

6. Overall, the torus of phases provides an excellent representation of the mindscape of a listener – at least to a fairly extensive corpus of western music. This last claim will be substantiated by the two next sections, see [96] for more.

It is important to understand the relationship between non-injectivity and voice-leading: remember that $\mathscr{F}_A(4)$ is invariant by minor-third transposition ($A \mapsto A+3$), just as $\mathscr{F}_A(3)$ is invariant by major-third transposition ($A \mapsto A+4$). This reflects well on the fact that (say) a seventh chord changes little when transposed by a minor third (resp. a triad by a major third): CEGBb turns into DbEbGBb, CEG to BEG♯ or even

---

are prevalent in the piece studied (cf. [15] again), or in octatonic pieces and in many actual analyses, see some examples in atonal music in Section 6.2.2.

[7] In Chopin's *Etude op. 10, N°5*, the right hand plays a pentatonic (black keys) and the left hand plays mostly the pitches of Gb major; those two scales have identical Fourier coefficients with odd indexes, which reflects spectacularly their compatibility.

two minor thirds, i.e. a tritone, turning CEGB♭ into C♯EF♯A♯(=B♭) (tritone transposition of sevenths being an idiomatic replacement in jazz music). These coefficients can be invariant by contextual inversions too. This vindicates the consideration of a_3 (resp. a_4) for 3 (resp. 4)-chords. More generally, it is well known that transposition of n/d of a d-set changes it little [91] if at all (in voice-leading distance terms), especially if the set is close to an even cyclotomy. This fact is well expressed by the conservation of the correspondent Fourier coefficient: proximity of phases goes hand in hand with short voice-leading distance. More specific and fruitful confusions are discussed in Section 6.2.2 below.

The major and minor triads in Fig. 6.3 appear to be aligned on the chromatic line of pcs alias diminished triads; actually, they are only very close to it, for reasons that will be elucidated in Section 6.3. The table of their coordinates, useful for analysis of tonal music, is provided in Table 8.34.

The disposition of dyads is somewhat messy, since major seconds and tritones do not have torus coordinates and

Lemma 6.7. *The minor third $(a, a+3)$ has the same phase coordinates as the semitone/minor second $(a-5, a-4)$.*[8]

Meaning that for instance EG and BC are the same point on the torus of phases.

Proof. Up to transposition (see Lemma 6.5) it is only necessary to verify the lemma for a specific value of a, say $a = 0$. But

$$\mathscr{F}_{\{0,3\}}(3) = 1 + e^{-3 \times 3.2i\pi/12} = 1 + e^{-18i\pi/12} = 1 + i$$
$$= e^{+15i2\pi/12} + e^{+12i2\pi/12} = \mathscr{F}_{\{-5,-4\}}(3).$$
$$\mathscr{F}_{\{0,3\}}(5) = 1 + e^{-3 \times 5.2i\pi/12} = 1 + e^{-30i\pi/12} = 1 - i \Rightarrow \varphi_5 = -\pi/4.$$
$$\mathscr{F}_{\{-5,-4\}}(5) = e^{+25i2\pi/12} + e^{+20i2\pi/12} = e^{2i\pi/12} - e^{4i\pi/12}$$
$$= e^{3i\pi/12}\left(e^{-i\pi/12} - e^{i\pi/12}\right) = -2i\sin\frac{\pi}{12}e^{i\pi/4} = 2\sin\frac{\pi}{12}e^{-i\pi/4} \Rightarrow \varphi_5 = -\pi/4 \text{ too.}$$

Notice that one Fourier coefficient is actually identical, whereas the other has a different magnitude but the same phase.

This apparent messiness is in fact instrumental in the disposition of triads, since these dyads are centers of symmetry between two adjacent triads as we will discuss in Section 6.3.

All of this means that a new musical space has been made available, which embeds the usual Tonnetz together with many (though not all) pc-sets with arbitrary cardinalities. For instance, this solves the conundrum of the proximity of C major and F minor triads: in a thoughtful comparison of distances ([91], pp. 412 and in more detail [90]) Tymoczko points out that though in voice-leading terms F minor looks and sounds closer to C major than, say, F major (and indeed FM → Fm → CM

---

[8] It follows immediately that the tetrachord $\{a-5, a-4, a, a+3\}$ e.g. the major seventh CEGB shares the same coordinates.

is a run-of-the-mill chord progression), in the Tonnetz, F major is two vertices away from C major (via A minor) while F minor is one step further still. This discrepancy vanishes in Fig. 6.3, together with another one pointed out by the same author: F minor and E♭ minor are the same distance (three steps) from C major though the first one shares one common tone with the latter; however, in Fig. 6.3, F minor with its common tone is closer. In the torus of phases, we can picture triads and (common) pcs and most pc-sets.[9] Quoting Yust again:

> ...there is a different way of topologically enriching the Tonnetz that pre-
> serves the musical insights [...] and leads to a concept of harmonic distance.
> Such mixing of different-cardinality sets is not possible in voice-leading
> spaces without forfeiting their basic geometric properties.

6.2.2 Phases between tonal or atonal music

Focusing on major and minor triads, I had blissfully ignored in the seminal [15] simpler collections such as single pcs or dyads. This major oversight was fortunately corrected by Yust in [96], who immediately made good use of single pcs in noticing that triads are positioned 'inside' the triangle whose vertices are the pcs of the triad.[10] For this book, I pick up an illuminating consequence of the coexistence of single pcs and triads, Yust's fine disambiguation of enharmony in the E major/F minor modulation in Schubert's String Quintet, Adagio movement, shown in Fig. 6.6.

In [96], he remarks that the 'slide' movement between triads would move E major= $\{4,7,11\}$ not to F minor= $\{5,7,12\}$, but to the enharmonic E♯ minor. Does this hair-splitting distinction really make sense? In the torus of phases, it does: clearly in Fig. 6.7 (borrowed from his paper and slightly modified), the downward movement through the common tone E is opposite to the slide transformation that would circle the torus the other way round, going 'up' on the picture (and coming back through its bottom).

---

[9] Ironically though, the augmented fifths that were suggested in Douthett and Steinbach's 'chickenwire' as the wormholes invisibly allowing shortcuts between triads are absent from this new and larger model.

[10] The quotation marks suggest taking the notion of 'inside' – on a torus! – with a grain of salt. More precisely, quoting Yust (ibid.):

> We can think of a pc-set's position in the space as the "average" of all the positions
> of its individual pcs, but with a cautionary note: because the space is a torus, as one
> pc gets further removed from the others in a particular dimension, its contribution to
> the "average" is attenuated. For instance, the notes C and G have a stronger influence
> on φ_5 of the C major triad because they are closer together, and similarly the notes C
> and E determine φ_3 more strongly. Therefore the position of C major [see Fig. 6.3]
> is not in the exact center of the triangle made by the individual pcs C, E, and G, but
> leaning towards the lower left side of it.

We will see however that a dyad is in the exact center of its two constituent pcs, and will not fail to remember that the notion of middlepoint is ambiguous in a quotient space, to say the least.

Fig. 6.6. A dramatic modulation in Schubert

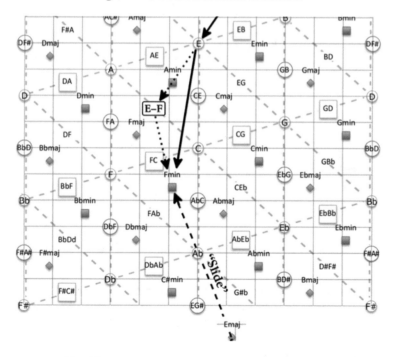

Fig. 6.7. Common tone modulation in the torus of phases

Similarly, distinctions between chromatic and diatonic movements which were lost in the Tonnetz make sense again in phase space.

Lastly, Yust introduces a promising visualisation tool, the *circle of diatonic scales*. Since a fifth transposition of a pc-set (here we choose a diatonic collection) rotates a_5 by $\pi/6$, focusing on the phase φ_5 *enables identification of the diatonic universe closest to the pc-set one is studying*, cf. Fig. 6.9 below where distinct voices of an atonal Stravinsky piece have a strong tonal/diatonic character.

In this vein, I include (Fig. 6.8) a simple but illuminating example of tracking a_5 in the first 11 bars of Mozart's *Sonata Facile* in C Major, K. 545 (bars 5-8 have been fused because they share the same pc-set, diatonic C scale). The last bars show quite well the half-heartedness of the modulation to G major (which Mozart undoes immediately afterwards in bars 12-13, trilling on a F♯ that descends to F natural), which can be inferred in the actual notes from the accidental C♯ and the absence of the leading tone F♯ in measure 9 sounding of D minor melodic, but even more clearly from the position of the a_5 arrow, between C and G scales (on the picture, letters index diatonic scales, e.g. C means 'diatonic C major")

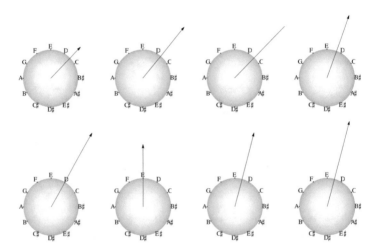

Fig. 6.8. The diatonic complex vector a_5 in Mozart K. 545's first bars

Still in the same vein, I devised a crude movie tracking the a_5 coefficient every 2 seconds of Berg's sonata.[11] Its size carries important information (the diatonic character), but its ever-changing direction reminds us that the composer does not want the listener to identify too clearly a stable tonal context. See
http://canonsrythmiques.free.fr/movies/bergSonata_a_5.gif

In the Stravinsky example already mentioned (see Fig. 4.19) it was argued that the a_5 vectors of the three instruments cancel each other out. It is illuminating to have a closer look at their individual phases. Again in Fig. 6.9, I represent these phases as angles on the unit circle, indexed by *the values of φ_5 for diatonic scales*.

---

[11] Personal MIDI recording.

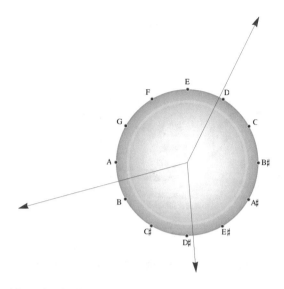

Fig. 6.9. a_5 for the three instruments in reference to *diatonic* scales

It is in accordance with musical perception that GABC appears close to C major, though verging towards G major – the two major scales it is a subset of – while C♯D♯EF♯ is right between E and B diatonic scales, etc. DFT analysis effectively subsums so-called 'complex analysis', where pc-sets are considered according to which archetypal supersets they are subsets of.

Quoting [100]:

> The diatonicity values also reflect the process of inferring a superset or a subset. When presented with a subset whose source scale is ambiguous, the φ_5 value essentially averages the values of possible supersets, with greater weight given to the most even possible 7-note supersets (especially possible diatonic supersets).

More generally, analysis of Fourier coefficients[12] both in magnitude and phase supersede complex analysis (i.e. inclusion relationships), fulfilling Allan Forte's dream better than previous, more naive tools. Consider a last drop of information in the Stravinsky excerpt: the similar diagram placing the three instruments viz. the three *octatonic* scales is just as illuminating, as we can see in Fig. 6.10; all three phases combine to strengthen the direction of the 01-octatonic scale, which is indeed the one closest in common tones/voice-leading to the union of all pc-sets.

---

[12] At least for a_4, a_5, a_6.

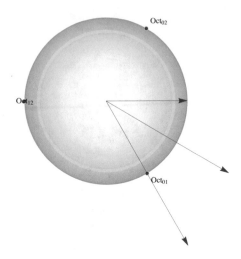

Fig. 6.10. a_4 for the three instruments viz. the *octatonic* scales

6.3 Central symmetry in the torus of phases

This last section expounds Yust's novel notion of local symmetries around parts of subsets, which generalise the LPR operations of Riemann transformational theory. I only developed a few calculations and theoretical points. Not only does this setting flesh out the abstract vertices and edges of the Riemannian Tonnetz, but since the torus of phases contains a near continuum of pc-(multi)sets besides pcs and triads, it provides hitherto unknown meaning to these transformations.

6.3.1 Linear embedding of the T/I group

In [15] I had remarked on the quasi-alignment of major and minor triads and concluded noting that diminished triads are squeezed in between. In [96], Yust understands the diminished triads as simply single pcs – according to the Tritone Lemma, BDF or $e25 = \{2, 5, 11\}$ has the same coordinates as just D or 2 – and further notices that triads are lined up according to successive symmetries around either a common pc or a common dyad, see below Fig. 6.17:

$$\ldots \quad 058 \quad 0 \quad 047 \quad 47 \quad e47 \quad e \ldots$$

In particular, the PLR inversions of neo-Riemannian theory appear as symmetries around common dyads. Other inversions like the well-discussed C major $\mapsto$ F minor are symmetries around one point (here C). See again Fig. 6.3.

Yust states a deceptively obvious lemma:

Proposition 6.8. *If A and B are symmetrical around a center c (resp. a dyad ab), then their torus projections are symmetrical around the torus image of c (resp. the image of the dyad).*

See for instance triads 045 and 945 in Fig. 6.11.

Proof. Consider two triads $A = \{x, y, c\}$ and $B = \{2c - x, 2c - y, c\}$ symmetrical around c. Then for any index k

$$\mathscr{F}_A(d) = e^{-2i\pi kx/n} + e^{-2i\pi ky/n} + e^{-2i\pi kc/n}$$

$$\mathscr{F}_B(d) = e^{-2i\pi k(2c-x)/n} + e^{-2i\pi k(2c-y)/n} + e^{-2i\pi kc/n}$$

$$= e^{-4i\pi kc/n} \left(e^{2i\pi kx/n} + e^{2i\pi ky/n} + e^{2i\pi kc/n} \right)$$

$$= e^{-4i\pi kc/n} \overline{\mathscr{F}_A(k)}$$

meaning that indeed $\dfrac{\arg(b_k) + \arg(a_k)}{2} = -\dfrac{2\pi kc}{n} = \arg(c_k)$, if a_k, b_k, c_k designate the kth Fourier coefficients of A, B and c respectively.

A similar proof works for a symmetry around a dyad ab, defined as $x \mapsto a + b - x$: $a + b$ plays the part of $2c$, i.e. the symmetry is around $\frac{a+b}{2}$, which is a pitch-class when $a + b$ is even and only virtual when $a + b$ is odd; the computation is otherwise identical.

On the face of it, this sounds like one can compute and draw midpoints directly on the torus of phases, which would mean that projection on the torus of phases would be an affine map.[13] But this would be ludicrous considering that

1. the maps involved (exp, arg) are non linear;
2. the space is not linear (a torus!), only locally planar;
3. the notion of midpoint itself is flawed: since $\varphi = \varphi + 2\pi$ for any angle, the 'middle' of (φ, ψ) and (φ', ψ') is not only $(\frac{\varphi + \varphi'}{2}, \frac{\psi + \psi'}{2})$ but also any of $(\pi + \frac{\varphi + \varphi'}{2}, \frac{\psi + \psi'}{2})$, $(\frac{\varphi + \varphi'}{2}, \pi + \frac{\psi + \psi'}{2})$ or $(\pi + \frac{\varphi + \varphi'}{2}, \pi + \frac{\psi + \psi'}{2})$!
4. and indeed this map is not always midpoint-preserving: for instance the midpoint of the torus images of 04 and 23 is $(1.1781, 0.3927)$ whereas their midpoint is the multiset $\frac{1}{2}(0234)$ whose image on the torus is $(0.7854, 0.6319)$.

The symmetries that are preserved through projection to the torus of phases are only the original T/I operations, not *all* symmetries on the space of distributions (e.g. characteristic maps of pc-sets). They constitute (together with translations) a group whose action on pc-sets is actually induced by its action on $\mathbb{Z}_n$. It is the T/I group of maps $x \mapsto a \pm x$, or if needed a continuous extension thereof:

T/I acts on $\mathbb{Z}_n$ as an extension of the group $\mathbb{Z}_n$, namely the semi-direct product $T/I = \mathbb{Z}_n \rtimes \mathbb{Z}_2$, and naturally induces an action on subsets of $\mathbb{Z}_n$ – or even on distributions, i.e. $\mathbb{C}^{\mathbb{Z}_n} \approx \mathbb{C}^n$. At this stage the immersion of $\mathbb{Z}_n$ in the continuous circle, considered as $\mathbb{R}/n\mathbb{Z}$, suggests introducing the continuous T/I, the group of maps $x \mapsto a \pm x$ with $a \in \mathbb{R}/n\mathbb{Z}$. We consider here a one-dimensional Lie group acting

---

[13] Any continuous map sending midpoint to midpoint preserves barycenter (starting with dyadic barycentric coordinates) and hence is affine. I am indebted again to [96] who cites [22] trying to construct such an affine space from pcs and their barycenters, but this model is ill-defined because pcs modulo octave are not compatible with barycentric operations.

on a n-dimensional (complex) manifold. This manifold projects on the 2D-torus of phases, and thus an action on this torus is induced. The result in Proposition 6.8 states that the continuous T/I is isomorphic[14] with the group of central symmetries and translations in the torus: for coefficients 3-5 and $n = 12$ we have the bijective correspondence between maps

$$T_A : (x \mapsto x+a) \mapsto \left(\begin{pmatrix} \varphi_3 \\ \varphi_5 \end{pmatrix} \mapsto \begin{pmatrix} \varphi_3 - a\pi/2 \\ \varphi_5 - 5a\pi/6 \end{pmatrix} \right)$$

$$T_A \circ I : (x \mapsto a-x) \mapsto \left(\begin{pmatrix} \varphi_3 \\ \varphi_5 \end{pmatrix} \mapsto \begin{pmatrix} -\varphi_3 - a\pi/2 \\ -\varphi_5 - 5a\pi/6 \end{pmatrix} \right)$$

mapping translations to translations and symmetries to symmetries. But the image group is still one-dimensional, a subgroup of the affine group on the torus. This works essentially because these transformations originate from single points (which share the same magnitude of Fourier coefficients): an element of $\mathbb{Z}_n$ generates both a translation and a central symmetry.

Another way to look at this is the following consequence of Proposition 6.8: the torus point for a dyad (a,b) is the same as it is for the middle-point $c = \dfrac{a+b}{2}$ of a and b (when it is an imaginary pc, say when $(a,b) = (0,3), c = 1.5$, then some Fourier coefficients may be undefined though not the third and fifth). Hence all single pcs and all dyads are naturally associated with one central symmetry on the torus.

But this is false for larger pc-sets, or even for the 'midpoint' of (a,b) and (c,d) as mentioned above (unless both dyads share the same shape). Another example: 045 has phase coordinates $(-0.463648, 0.261799)$ but the center of this triad[15], namely 3, has coordinates $(\pi/2, -\pi/2) \approx (1.57, -1.57)$ which put it completely 'outside' the 'triangle'. Neither does the triad 045 stand dead center of the triangle (see Fig. 6.11). Yust's symmetry appears however, when one looks at 459, symmetrical of 045 around the midpoint of the images of 4 and 5 which is the image of the dyad 45. This can be understood as a consequence of the diminution of dimension, from the set of all distributions to a two-dimensional space.

Even the intuitive notion of 'triangle' is blown up in this model: in Fig. 6.12 I have drawn the 'straight lines'[16] passing through C, E and G, i.e. (0, 4, 7) on a 3D-rendering of the torus, and we can notice with some surprise that the CG edge passes through all pcs (it is the cycle of fifths: starting low right with C, the next pc is G, then the red line goes through D and A before reaching E)[17] so that the vertices of

---

[14] Essentially because of two things: taking the phase of a Fourier coefficient of a pc-set basically turns addition in $\mathbb{Z}_n$ to multiplication through complex exponentiation and back to addition, and the actions of translations and central symmetries on pc-sets are point-wise.

[15] Notice that already in $\mathbb{Z}_{12}$, a triad, collection of 3 pcs *modulo 12*, has nine distinct centers, just as a dyad has two midpoints a tritone away. Yust noted *supra* that this works approximately for major/minor triads but breaks down when moving away; this is because the arg map, being differentiable, is locally linear.

[16] Technically, geodesics.

[17] On the other hand, the horizontal circle joining C and E only connects G♯, and the last edge of the triangle DG or (47) includes the 4 pcs of the diminished seventh that contains it.

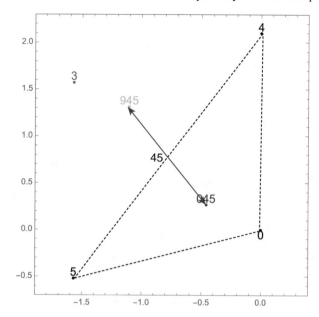

Fig. 6.11. The barycenter 3 of 045, pc-wise, is off center

the triangle are in fact aligned! Indeed, *there are infinitely many straight lines that join all pcs*. This bodes ill for the notion of center of the triangle, and indeed there are 9 legitimate candidates, including the one close to the position of the 047 triad. They appear as stars on the picture; three of them lie on the cycle of fifths.

6.3.2 Topological implications

This allows extremely smooth moves (and short ones too) in some progressions of chords. We have already seen that the PLR operations of neo-Riemannian harmony occur in the torus of phases between adjacent triads. Moreover, the central symmetries in Proposition 6.8 place firmly as a middle point (say) a dyad CE between related/symmetrical triads CEG and ACE, or a point/diminished triad C alias ACE♭, between CEG and the symmetric FA♭C. This locally convex geometry, together with the capacity of mapping most of the possible pc-(multi)sets on the torus, leaves room for very smooth paths indeed between different chords. There is a suggestion (that could be formalised by using arbitrary multisets) of tending towards the continuous torus underneath, which is reminiscent of quantisation in Chapter 5.

   Yust explored in [97] the generalisation of the zig-zag between major/minor triads by central symmetries around a common tone or a common dyad. He starts from the famous PLR operations, three such symmetries around a given major or minor triad, see Fig. 6.13, which leave a dyad invariant.

   For a more generic example, consider the good voice-leading obtained by flipping a dominant seventh (say GBDF or $\{11,2,5,7\}$) around its tritone (not to con-

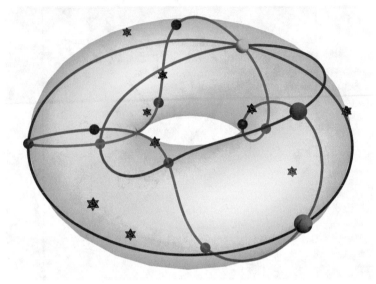

Fig. 6.12. The complete 'triangle' 047 on the torus

fuse with the more jazz-idiomatic tritone transposition) 11-5, yielding $\{11, 2, 5, 9\}$ or

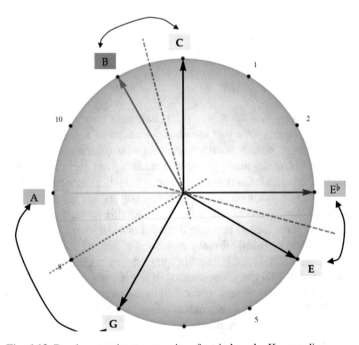

Fig. 6.13. Dyad-preserving symmetries of a triad on the Kremer diagram

BDFA, a minor seventh (or Tristan chord). The quality of the voice-leading is only to be expected, since

- Already a dyad is fixed;
- other pc-sets may be fixed too (here D);
- pc-sets which are actually moved have fewer places to go because of all the ones we already have taken into account.

These movements can be easily expressed in the torus of phases thanks to the formula in Lemma 6.5. This kind of voice-leading-producing symmetry is very general; one only has to partition a pc-set into two sub-pc-sets, both of which admit a central symmetry. It is then possible to concatenate symmetries between the different parts in a nice flip-flop movement. For instance, we turned GBDF into BDFA with the operator $x \mapsto 4 - x$. Applying it again would of course restore GBDF, but there is a new symmetry available, $x \mapsto 11 - x$ which exchanges D-A and turns the whole chord into another dominant seventh, $\{0, 2, 6, 9\}$ or CDF#A. This can be carried further in a nice cyclic progression which generalises the cycles on the Tonnetz, called by Yust a 'Tonlinie'.[18] Fig. 6.14 shows such a progression between 12 dominant sevenths and Tristan chords.

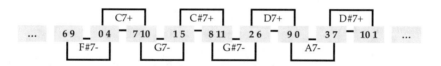

Fig. 6.14. A 'Tonlinie' with all dominant sevenths and their inversions in the torus of phases

It should also be noticed that the most famous example of symmetry between a Tristan chord and a dominant seventh – repeated several hundreds of times in the eponymic opera – uses a different symmetry, one leaving invariant a minor third instead of the common tritone. Including the symmetry above, there are three ways to generate the chimeras listed in Fig. 6.15, keeping the ambient pc-set in the shape of a reunion of two chromatic tetrachords which sustain the melodic content, cf. [5]. Since there are three decompositions of a 4-chord into pairs of (self-symmetric) dyads, there are in all six ways to move from a dominant seventh to a Tristan chord by symmetry around a common dyad, providing in phase-space a picture not dissimilar to the usual Tonnetz and its hexagonal symmetries.[19]

In general one may state the following:

---

[18] This generalisation was what Yust aimed at initially. But generically, the chain obtained has fewer symmetries than the whole dihedral group, symmetry group of the Tonnetz.

[19] There is actually a fourth possible decomposition type – turning say (0 4 7 10) to (0 2 5 8) – whose elucidation we leave to the reader, with the additional exercise of composing another variant of Tristan's motif connecting these two chords.

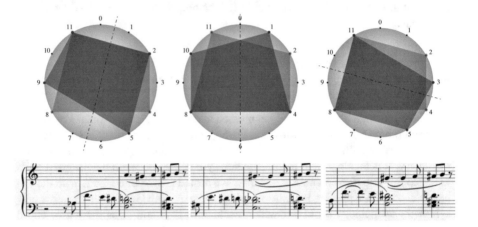

Fig. 6.15. Three different flip-flops between Tristan chord and 7+

Proposition 6.9. *(Yust 2014)*

Assume that $X = A \cup B$ with $A \cap B = \varnothing$ and that A (resp. B) is invariant under the axial symmetry s_A (resp. s_B).

Then $X' = A \cup s_A(B)$ and $X'' = s_B(A) \cup B$ are inversions of X (and a fortiori homometric with X).

Proof. Obvious since for instance $X' = s_A(A) \cup s_A(B) = s_A(A \cup B) = s_A(X)$. Also notice that X' and X'' are transpositions one of another, since $X'' = s_B(s_A(X'))$ and a composition of two inversions is a transposition.

When is such a flip-flop construction available? I clarified the possibilities with the following:

Proposition 6.10 (Amiot 2014).

Any pc-set in $\mathbb{Z}_{12}$ can be decomposed in such a way.

Proof. Computer made with disjunction of cases, building k-subsets from symmetrical subsets and comparing the number of solutions with the expected $\binom{12}{k}$. For instance there are 52 symmetrical 3-sets (arithmetic sequences all, e.g. 012 or 024 and so on), and disjunct unions of these with the 66 dyads yields all 792 5-sets in $\mathbb{Z}_{12}$. On the other hand, some hexachords cannot be decomposed into two trichords, each symmetrical, but 4+2 decompositions yield all hexachords which are thus amenable to flip-flop sequences (118 out of 924 can be decomposed either as 5+1, 4+2 or 3+3. An example is $\{0,1,3,5,6,9\}$; the whole gamut of these maximally symmetrical hexachords (up to transposition), which may be of interest to composers, is provided in Table 8.35).

This should not be overdone: in larger chromatic universes the proposition fails, since

Proposition 6.11. *The set $\{0,1,3,8,12\}$ cannot be partitioned in two symmetrical subsets in $\mathbb{Z}_n$, for $n \geqslant 17$.*

Proof. See Fig. 6.16: none of the ten 3-subsets or five 4-subsets is symmetrical.

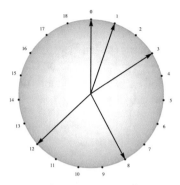

Fig. 6.16. Subset $\{0,1,3,8,12\}$ in $\mathbb{Z}_{19}$

6.3.3 Explanation of the quasi-alignment of major and minor triads

It is such a chain of central symmetries that runs through major and minor triads. On the lower image in Fig. 6.17, the slopes are exaggerated the better to distinguish minor, major and diminished triads which appear all but aligned on the upper part of the picture.

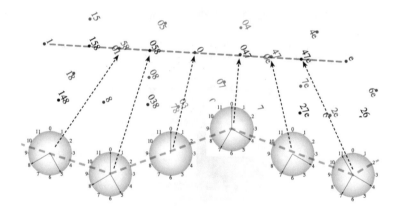

Fig. 6.17. Triads are on a zig-zag very close to a straight line

Primes (coincident with diminished triads as we have seen) are genuinely aligned with minor thirds: the phase coordinates of 11, 0, and 1 are $\left(\frac{\pi}{2},\frac{5\pi}{6}\right),(0,0),\left(\frac{\pi}{2},\frac{5\pi}{6}\right)$ whereas for (3 6), (4 7), (5 8) they are $\left(\frac{3\pi}{4},\frac{5\pi}{4}\right),\left(\frac{\pi}{4},\frac{5\pi}{12}\right),\left(-\frac{\pi}{4},\frac{5\pi}{12}\right)$ which all lie on the chromatic line $5\varphi_3 = 3\varphi_5$. The presence of coefficients 3 and 5, the indexes of the relevant Fourier coefficients, is no coincidence but a consequence of Lemma 6.5.

The reason why the two parallel lines of major and minor triads are so close in Fig. 6.17 is the closeness of the angles between the zigs and the zags, or equivalently between the line of centers and the direction of a triad from one of those centers:

namely, the line joining chromatically consecutive single pcs (or minor thirds) has a slope of $5/3$ as we have seen above, whereas the Fourier coefficients of 047 are $(2+i, 1+e^{i\pi/6}+e^{2i\pi/3})$ with phases $(\arctan\frac{1}{2}, \frac{\pi}{4})$, and hence the slope of the segment between 0 and 047 is $\frac{\pi}{4\arctan(1/2)} \approx 1.69395$, to be compared with its close rational approximation $5/3 = 1.666666\ldots$ which is the slope of the true chromatic line from which the triad departs ever so slightly.

Exercises

Exercise 6.12. Compute the variation of a_2 induced by a semitone move of a single pc (say C$\mapsto$C$\sharp$).

Exercise 6.13. Compute the variation of a_3, a_5 and φ_3, φ_5 between the C major triad $\{0,4,7\}$ and one of its three neighbours in the Tonnetz, $\{11,4,7\}, \{0,3,7\}$ or $\{0,4,9\}$. Use Table 8.34.

Exercise 6.14. Compute (φ_3, φ_5) for the chord CDEA$\flat$.

Exercise 6.15. Label all pcs in Fig. 6.12.

Exercise 6.16. Compose a variant of Tristan's motif connecting the two chords $\{0,4,7,10\}$ and $\{0,2,5,8\}$. What simple geometrical transformation takes one pc-set into the other?

Exercise 6.17. Check Proposition 6.10 on a couple of pc-sets (at least hexachords).

Exercise 6.18. Check all trichords in Fig. 6.16 for symmetries.

7

Conclusion

The use of DFT in music theory really soared after the notion was resuscitated from Lewin's work by Quinn [72] in 2005. As we have seen, it bears the tremendous advantage that each coefficient, and moreover each polar coordinate of each coefficient, yields dramatically important musical information (say, the phase of a_5 shows which diatonic universe is closest to the pc-set in question). Some musical qualities are immediately visible in Fourier space whereas they require computations in the original musical domain (say, pc-distributions); Fourier space, with this minimised computational complexity, is closest to our perception of music. Indeed, psycho-acoustic experiments on the perception of saliency (and its evil twin, low saliency including nullity of a coefficient) should be enhanced and furthered, since Fourier qualities seem to mirror exactly musical features processed by the human brain.

Theorem 1.11 shows that Fourier space is unique in that respect. Consequently, DFT analysis should be expected to rise as one of the most useful tools in music theory. The present book introduces the state of the art on the subject; it is also intended as a textbook for future work, both in setting down clear and comprehensive definitions and properties (including the alternative versions of DFT in continuous spaces) and suggesting, through simple examples, some ways of using DFT for practical work. The neo-classical theory of consonance [79] – coincidence of harmonics – probably bears points of convergence with DFT, which cry out for exploration.

I hope that Chapter 2 will stimulate compositional creativity using homometric pc-sets or rhythms. Though theoretically solved for real-valued distributions, the homometry problem (find all subsets homometric to a given one) is still open even for rational values, though the difficult Theorem 2.10 opens new routes for practical exploration. This may be a good topic through which to explore the recurring, puzzling question of singling out integer (or even 0-1) solutions to a problem that is already solved in the real or rational fields. Exploration of the continuous orbit of homometric real-multisets and integer approximation through linear- or constraint-programming is another promising tack. Moreover, the status of the inversions, already identified as local operators in various theoretical works, is enriched by the present study where their order (in the group of all transformations) appears to be infinite. The music-

© Springer International Publishing Switzerland 2016
E. Amiot, *Music Through Fourier Space*, Computational Music Science,
DOI 10.1007/978-3-319-45581-5_7

theoretical implications of this phenomenon are still to be fathomed. Working in the real field is of course natural in continuous spaces, and the 'DJ' approach of Noll or Beauguitte – smoothly moving one Fourier coefficient, thus changing the whole distribution in real space – has a lot of potential.

The question of musical (mostly rhythmic) tilings also deserves greater recognition, not only because of the numerous possible compositional or analytic applications, but also as one of the few topics where musical notions pave the way to solutions to hard mathematical problems. At least one aesthetic aspect of this topic is virgin territory: as we have explored in Chapter 4, the nullity of certain Fourier coefficients which is essential to tiling actually means that some musical characters are absent in the subset (for instance, since $a_5 = 0$ for a diminished seventh, this chord is not diatonic; more generally, any tiling pc-set has some 0 coefficients and hence must miss some characters). This opens up fascinating possibilities[1] of choosing and even cataloging rhythms according to the presence or absence of these characters, linked to periodicities, generatedness, evenness or saturation in some sub-intervals, as proved in Chapter 4.

The very notion of saliency was justly put to the fore by Quinn. Later work (mostly Yust's brilliant musical analyses) have shown how neatly and effortlessly the size of Fourier coefficients helps to discriminate between essential characters in tonal or atonal music, e.g. diatonic, octatonic, chromatic, 'quartal' and so on. It is an exhilarating perspective finally to be able to tell without doubt whether early 20th-century Slavic music is truly octatonic (see Fig. 7.1), or to measure diatonicity in Strauss' *Elektra* or *Salome* with precision. I have suggested the momentous route of studying the six characters for hexachords/tropes in serial dodecaphonic music; of course, similar analysis would be even more appropriate for music like Matthias Hauer's, who uses tropes as pc-sets while the second Viennese school more or less clings to ordered hexachordal sequences. This sequenciality`is very precisely addressed by Noll's ordered DFT $\mathfrak{F}_{\mathscr{A}}$, which undoubtedly deserves better recognition and further developments.

On the other hand, I reckon that 'cancellations of diatonicity' such as Yust discovered in Stravinsky (and Debussy or Satie, among others, not to mention my own Berg examples) bring up hitherto unavailable levels of understanding of music on the brink of tonality. These cancellations can only be explained by the consideration of phase – directions of Fourier coefficients – which seems a promising field of exploration too, even in tonal music as can already be seen on my simplistic Mozart example (Fig. 6.8). Paraphrasing Yust, phase spaces are a clear improvement on ambiguous 'complex' set-theoretic analysis. Moreover, his own discovery of 'split symmetries' generating zig-zags of pc-sets opens up new chapters for analysis and musical composition. Indeed, he discovered that even when restricted to saliency, DFT includes and supersedes classical 'set theory', and in a way that seamlessly includes the classical Tonnetz, thus enhancing our understanding of this model hitherto *unyustly* restricted to major/minor triads.

---

[1] Most of this book addresses pc-sets in $\mathbb{Z}_{12}$, but working with a larger n, be it in the rhythmic domain or with microtonality, greatly enriches perspectives.

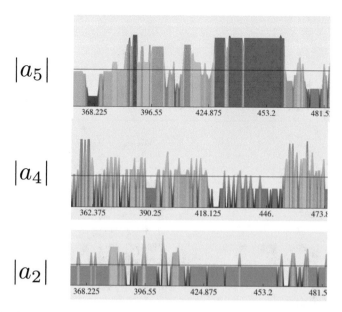

$|a_5|$

368.225   396.55   424.875   453.2   481.5

$|a_4|$

362.375   390.25   418.125   446.   473.8

$|a_2|$

368.225   396.55   424.875   453.2   481.5

Fig. 7.1. Diatonic, octatonic, quartic character in a few bars of Stravinsky's *Rite of Spring*

Saliency and nullity of coefficients are apparently opposite perspectives in DFT analysis. However, because of Theorem 1.8, the higher some coefficients soar, the lower the other coefficients must be. In the same vein but more extreme, Heisenberg's inequality (Theorem 4.46) showed that nullity of Fourier coefficients is limited by the number of 0's in the distribution. We have seen how the same tools are effective in both cases, as well as in the middle case exemplified by difference sets and FLID. In particular, it is imperative to explore thoroughly the possible transformations between subsets with similar characters, exact or approximate – homometric sets, all complements to a given tiling motif, affine orbits of FLID, diatonic (or augmented/octatonic etc.) pc-sets, and so on. We have seen a few possibilities in this direction, moving between and transforming sets with similar Fourier flavours. This calls for the development of easily available and user-friendly computer tools. Those that I developed alongside this book are only intended as feelers into this open journey.

A new continent is open for exploration. I have been fortunate to be among the first to set foot on it; as can be inferred from the perusal of the last 200 pages, those strolling its shores only have to stoop to gather precious gems. I hope that many musicians, 'pure' mathematicians, composers and theorists, will be enticed to join us in farming its innumerable riches.

8

Annexes and Tables

Summary. This chapter features solutions to selected exercises, some pictures chosen from the online database of all profiles of pc-sets
`http://canonsrythmiques.free.fr/MaRecherche/photos-2/`
which have been included here because they are mentioned in the main text, and, for reference, tables of singular pc-sets, phases of triads, enumeration of the most symmetrically pc-sets in the sense of Proposition 6.10, and values of Major Scale Similarity for a large panel of historical temperaments.

8.1 Solutions to some exercises

1.39 All sums run over the whole $\mathbb{Z}_n$:

$$\widehat{f * g}(x) = \sum_k (f * g(k)) e^{-2i\pi kx/n} = \sum_k \sum_j f(k-j)g(j) e^{-2i\pi(k-j+j)x/n}$$

$$= \sum_k \sum_j f(k-j)g(j) e^{-2i\pi(k-j)x/n} = \sum_k f(k-j) e^{-2i\pi(k-j)x/n} \times \sum_j g(j) e^{-2i\pi jx/n}$$

$$= \sum_\ell f(\ell) e^{-2i\pi \ell x/n} \times \sum_j g(j) e^{-2i\pi jx/n} = \widehat{f}(x) \times \widehat{g}(x).$$

1.41 We have

$$\mathscr{F}_{A+p}(x) = \sum_{k \in (A+p)} e^{-2i\pi kx/n} = \sum_{\ell \in A} e^{-2i\pi(\ell+p)x/n}$$

$$= e^{-2i\pi px/n} \sum_{\ell \in A} e^{-2i\pi \ell x/n} = e^{-2i\pi px/n} \mathscr{F}_A(x).$$

If $\mathscr{F}_A(x) \neq 0$ this yields $e^{-2i\pi px/n} = 1$, i.e. $px/n \in \mathbb{Z}$.

2.38 Fig. 8.1 is an excerpt of a small composition.

2.41 The direct part uses the convolution product:

© Springer International Publishing Switzerland 2016
E. Amiot, *Music Through Fourier Space*, Computational Music Science,
DOI 10.1007/978-3-319-45581-5_8

Fig. 8.1. The two hands play reverse intervals in two Z-related pc-sets

$$h = \frac{2}{n}(1,1\ldots 1) - (1,0,0\ldots) = \frac{2}{n}\mathbf{1} - \delta.$$

Consider any hexachord A and its characteristic map $\mathbf{1}_A$:

$$h * \mathbf{1}_A = \frac{2}{n}\mathbf{1} * \mathbf{1}_A - \delta * \mathbf{1}_A = \frac{2\#A}{n}\mathbf{1} - \mathbf{1}_A = 1 - \mathbf{1}_A$$

when $2\#A = n$, i.e. A is a generalised hexachord (it divides $\mathbb{Z}_n$ in two parts of same size), and the map that we computed is 0 when $x \in A$ and 1 else, i.e. $h * \mathbf{1}_A$ is equal to the characteristic function $\mathbf{1}_{\mathbb{Z}_n \setminus A}$ of the complement of A.

To prove that h is a spectral unit we must study its eigenvalues. The matrix $\mathcal{H}$ derives from the matrix $\mathbf{1}$ with only ones, whose nullspace has dimension $n-1$ (the hyperplane $x_1 + \ldots x_n = 1$) and hence 0 is an eigenvalue with multiplicity $n-1$. The other eigenvalue is n, associated with vector $(1,1\ldots 1)$. Hence the eigenvalues of $\mathcal{H}$ are

$$\frac{2}{n} \times 0 - 1 = -1 \qquad \frac{2}{n} \times n - 1 = 1.$$

Both eigenvalues have magnitude 1; we have proved that h is a spectral unit, connecting any hexachord and its complement.

2.42 The Fourier coefficients of the spectral unit $j^3 = (0,0,0,1,0,0,0,0,0,0,0,0)$ (i.e. the eigenvalues of its matrix) are $(1,-i,-1,i,1,-i,-1,i,1,-i,-1,i)$. Choosing arbitrarily cubic roots of each of the 12 coefficients yields cubic roots of j^3, but most of these $531,441$ distributions are irrational. One example (choosing the smallest phases for all cubic roots) is

$$\left(\frac{3}{8} + \left(\frac{1}{4} + \frac{i}{8}\right)\sqrt{3}, 0, 0, \frac{3}{8} - \frac{i\sqrt{3}}{8}, 0, 0, \frac{3}{8} - \left(\frac{1}{4} - \frac{i}{8}\right)\sqrt{3}, 0, 0, -\frac{1}{8} - \frac{i}{\sqrt{3}}8, 0, 0\right).$$

To ensure *rational* spectral units, we must use Thm. 2.10, which determines all coefficients from the ξ_d, $d \mid n$. From it we get that $\xi_0 = +1$ (the case -1 is impossible), that for $d \in \{1,2,3,6\}$, ξ_d is any power of $e^{2id\pi/12}$, and ξ_4 is a power of $e^{4i\pi/12}$; lastly, for any k coprime with 12, $\xi_{kd} = \xi_d^k$ (happily or by design, the last, complicated case will not occur in this exercise).

Since $\xi_1^3 = -i = e^{3i\pi/2}$ we have three choices: $\xi_1 \in \{e^{i\pi/2}, e^{7i\pi/6}, e^{11i\pi/6}\}$. The corresponding values of ξ_5, ξ_7, ξ_{11} are then determined (for instance $\xi_5 = \xi_1^5$). Similarly $\xi_2 \in \{-1, e^{5i\pi/3}, e^{i\pi/3}\}$ and hence $\xi_{10} = \overline{\xi}_2$.

The constraint of the theorem appears for ξ_3 which must be a power of i. The only possibility is then $\xi_3 = -i$ and $\xi_9 = i$; $\xi_4 \in \{1, j, j^2\} = \{1, e^{2i\pi/3}, e^{4i\pi/3}\}$ hence ξ_8. Lastly $\xi_6 = -1$ and we are reduced to $3^3 = 27$ solutions, which can be produced by inverse DFT or matrix products (the amount of computation is the same).

A typical rational cubic root of j^3 is

$$\left(-\frac{1}{4}, \frac{1}{4}, 0, \frac{1}{4}, \frac{1}{4}, \frac{1}{2}, -\frac{1}{4}, \frac{1}{4}, 0, \frac{1}{4}, \frac{1}{4}, -\frac{1}{2}\right).$$

3.66 $\Phi_1 = X - 1, \Phi_2 = X + 1, \Phi_3 = X^2 + X + 1, \Phi_4 = X^2 + 1, \Phi_6 = X^2 - X + 1, \Phi_{12} = X^4 - X^2 + 1$.

3.67 $\Phi_{16}(X) = \dfrac{X^{16} - 1}{X^8 - 1} = X^8 + 1$.

3.69 Singular: by rote, there are as many odd and even elements, so $a_6 = 0$.

3.70 (CG) is the sum of all 6 fifths beginning on C♯, D♯, F, G, A, and B, minus the 5 fifths beginning on D, E, F♯ G♯ and A♯.

3.73 One could compute $A(e^{2i\pi/30})$ numerically but this is not a rigorous proof (trigonometric computation is possible but deep). Best is to check that $A(X)$ is divisible by $\Phi_{30}(X) = X^8 + X^7 - X^5 - X^4 - X^3 + X + 1$. Polynomial division yields quotient $X^{16} - X^{15} + X^{14} + X^{11} + X^9 + X^7 + X^5 + X^3 + 1$ and remainder 0.

3.76 $A(X) = (1 + X^5)(1 + X^8) = \dfrac{X^{10} - 1}{X^5 - 1}(1 + X^8) = \Phi_{10}\Phi_2\Phi_{16}$.
Hence $R_A = \{2, 10, 16\}$ (A tiles $\mathbb{Z}_{16}$).

4.56 Try multiplying the first line $(0, 1, 3, 8, 12, 18)$ by $2, 4, 8\ldots$ Because of multipliers, there are three affine maps transforming each voice into another given one.

5.12 The smallest values of q satisfying the formula for $N = 100$ (i.e. both rational approximations are closer than $1/(10q)$) are $q = 36, 63, 70, 99\ldots$. For instance $\left(\dfrac{140}{99}, \dfrac{311}{99}\right) - \left(\sqrt{2}, \pi\right) \approx -(0.0000721, 0.000179)$, both coordinates well under $1/990$.

5.16 The proof follows the same pattern as the case developed in the example. Let $A = \{a_1, \ldots a_d\}$ be a d-subset of $\mathbb{Z}_n$ (or indeed of the continuous circle modulo n) and $B = \{b_1, \ldots b_d\}$ be a subset of a regular d-polygon, i.e. $d(b_i - b_j) \in \mathbb{Z}\ \forall i, j$, which is equivalent to $|\mathscr{F}_B(d)| = d$ as we have seen previously.

Assume B is the closest to A among similar subsets. Then by derivation

$$\frac{dAB^2}{db} = 2\sum(b_k - a_k) = 0$$

where db stands for any db_i since they are differentially identical. If B is written as a particular type of subset of a polygon, e.g. $B = x + \{\ldots b_0 + \dfrac{m_k}{d} \ldots\}$ with a specific distribution of the integers m_k, this pinpoints the value of the offset x (modulo n/d)

and hence of B, but this is not relevant in the following computation, insofar as we can assume that the quantities $b_k - a_k$ are small. We now compute the Fourier coefficient $\mathscr{F}_A(d)$:

$$\mathscr{F}_A(d) = \sum_k e^{-2i\pi da_k/n} = \sum_k e^{-2i\pi d\left(b_k+(a_k-b_k)\right)/n} = e^{-2i\pi db/n} \sum_k e^{-2i\pi d(a_k-b_k)/n}$$

where b stands for any b_k, since $e^{-2i\pi db_k/n}$ is independent of k by our assumption on the geometry of B.

Putting $\varphi_k = a_k - b_k$ one gets

$$|\mathscr{F}_A(d)| = \left|\sum_k e^{-2i\pi d/n\varphi_k}\right| = \left|\sum_k \left(1 - 2\frac{\pi d}{n}\varphi_k - \frac{2\pi^2 d^2}{n^2}\varphi_k^2 + \dots\right)\right| \approx d - \frac{2\pi^2 d^2}{n^2}\sum_k \varphi_k^2$$

since $\sum \varphi_k = 0$. This yields the formula since $\sum_k \varphi_k^2 = VL^2$.

6.12

$$e^{-2i2\pi/12} - e^0 = e^{-i\pi/3}\left(e^{-i\pi/3} - e^{+i\pi/3}\right) = -2i\sin\frac{\pi}{3}e^{-i\pi/3}$$

$$= 2\sin\frac{\pi}{3}e^{-i\pi/3-i\pi/2} = \sqrt{3}e^{-5i\pi/3/6}.$$

6.13 Between CEG and BDG the change is the same as between C and B. For a_3 it is $\Delta a_3 = e^{-2i3\times11\pi/12} - e^0 = i - 1 = \sqrt{2}e^{3i\pi/4}$.

However the phase of CEG is $\varphi_3 = 0.46365$ and for BEG it is 1.10715; hence the variation of phase is $\Delta\varphi_3 = 1.10715 - 0.46365 = 0.64350$. Similarly we find

$$\Delta a_5 = e^{-2i5\times11\pi/12} - e^0 = e^{10i\pi/12} - 1 = 2\sin\frac{5\pi}{12}e^{5i\pi/12}$$

and $\Delta\varphi_5 = 1.83260 - 0.78540 = 1.04720$.

Notice that the phase of the difference Δa_k is not the difference of phases $\Delta\varphi_k$: the first is the direction of a vector in Hoffman's space $(\arg(a_k - b_k))$ and the second a difference of one coordinate in the torus of phases $(\arg(a_k) - \arg(b_k))$. This illustrates the fact that the map arg is not linear.

6.14 $a_3 = 2$ and $a_5 = \frac{1}{2} + \frac{i\sqrt{3}}{2}$ so $\varphi_3 = 0, \varphi_5 = \pi/3$.

6.16 The inversion around 0 turns $\{0,4,7,10\}$ into $\{0,2,5,8\}$ (central symmetry $x \mapsto 12 - x$). A tentative motif between the minor seventh and the dominant seventh is given in Fig. 8.2 (I will readily agree that Wagner's version is better).

Fig. 8.2. Another Tristan chimera

8.2 Lewin's 'special cases'

Fig. 8.3. Table of all classes of singular pc-sets

8.3 Some pc-sets profiles

Fig. 8.4. Second/seventh

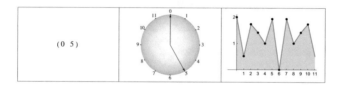

Fig. 8.5. Fourth/fifth

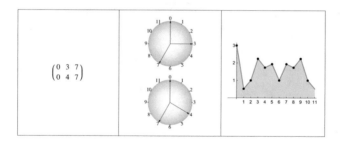

Fig. 8.6. Major/minor triad

Fig. 8.7. Rock/blues bass

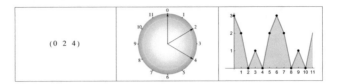

Fig. 8.8. Whole-tone trichord

Fig. 8.9. Chromatic trichord

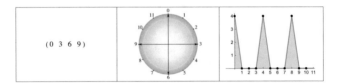

Fig. 8.10. Diminished seventh

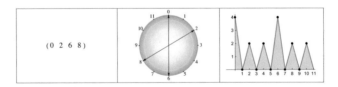

Fig. 8.11. Chunk of whole-tone scale

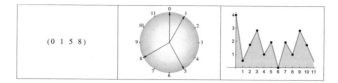

Fig. 8.12. S.N.C.F. jingle

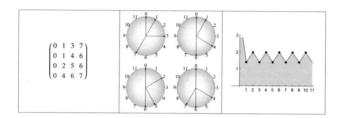

Fig. 8.13. Homometric quadruplet

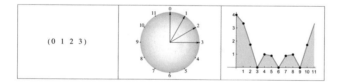

Fig. 8.14. Chromatic tetrachord

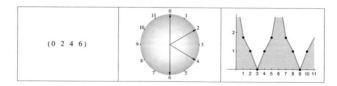

Fig. 8.15. Whole-tone tetrachord

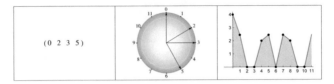

Fig. 8.16. An octa/diatonic tetrachord

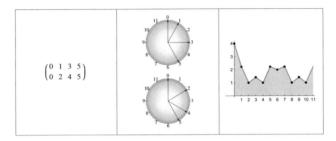

Fig. 8.17. A rather diatonic tetrachord

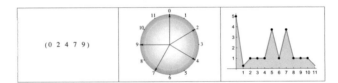

Fig. 8.18. Pentatonic scale

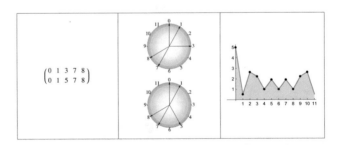

Fig. 8.19. Beginning of *La Puerta del Vino*

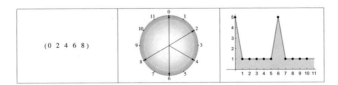

Fig. 8.20. Whole-tone pentachord

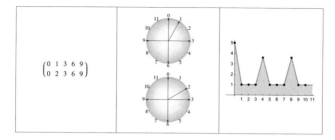

Fig. 8.21. A pentachord saturated in minor thirds

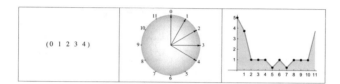

Fig. 8.22. Chromatic pentachord

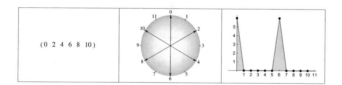

Fig. 8.23. Whole-tone scale

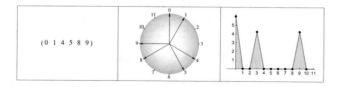

Fig. 8.24. Magic hexachord

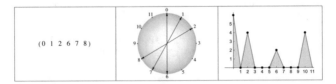

Fig. 8.25. Messiaen Mode M5

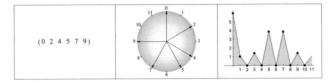

Fig. 8.26. Guidonian hexachord

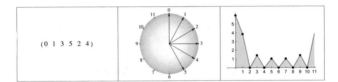

Fig. 8.27. Chromatic hexachord

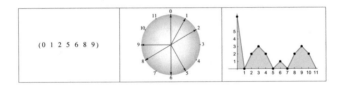

Fig. 8.28. Balanced seven-note scale

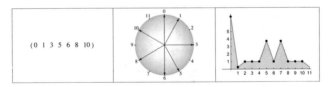

Fig. 8.29. Diatonic scale

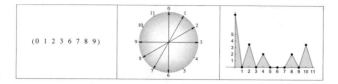

Fig. 8.30. Messiaen Mode M4

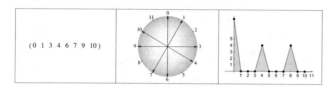

Fig. 8.31. Octatonic scale or M2

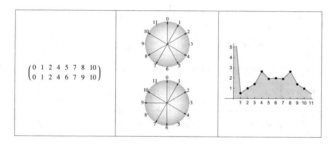

Fig. 8.32. An 'octatonish' collection in Stravinsky

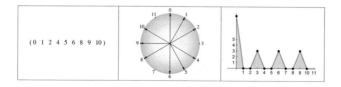

Fig. 8.33. Nonatonic scale or Messiaen Mode M3

8.4 Phases of major/minor triads

triad		θ_3	θ_5	triad		θ_3	θ_5
047		0,46365	0,78540	2611		2,67795	2,35619
058		-0,46365	-0,78540	1610		-2,67795	-2,35619
158		-1,10715	-1,83260	037		1,10715	-0,26180
169		-2,03444	2,87979	2711		2,03444	1,30900
269		-2,67795	1,83260	038		0,46365	-1,30900
2710		2,67795	0,26180	148		-0,46365	-2,87979
3710		2,03444	-0,78540	149		-1,10715	2,35619
3811		1,10715	-2,35619	259		-2,03444	0,78540
049		-0,46365	1,30900	2510		-2,67795	-0,26180
4811	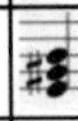	0,46365	2,87979	3610	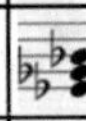	2,67795	-1,83260
1510		-2,03444	-1,30900	3611		2,03444	-2,87979
059		-1,10715	0,26180	4711		1,10715	1,83260

Fig. 8.34. Phase coordinates of major and minor triads

8.5 Very symmetrically decomposable hexachords

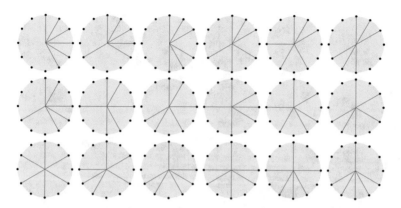

Fig. 8.35. The 18 most decomposable hexachords (up to transposition)

8.6 Major Scales Similarity

	MSS:	F	F♯	G	G♯	A	A♯	B	C	C♯	D	D♯	E
Zarlino	59	0	112	204	316	386	498	590	702	814	884	1 017	1 088
MeanTone15	80	0	114	195	308	389	503	616	697	811	892	1 005	1 086
MeanTone16	117	0	110	196	306	392	502	612	698	807	894	1 004	1 090
WM2	120	0	82	196	294	392	498	588	694	784	890	1 004	1 086
Pythagore	142	0	114	204	294	408	498	612	702	816	906	996	1 110
Kirnberger2	147	0	90	204	294	386	498	590	702	792	895	996	1 088
Kirnberger3	164	0	90	195	294	386	498	590	698	792	890	996	1 088
Vallotti	164	0	94	196	298	392	502	592	698	796	894	1 000	1 090
WM1	181	0	90	192	294	390	498	588	696	792	888	996	1 092
Lindley94	224	0	108	200	305	402	502	606	699	807	901	1 004	1 104
WM3	235	0	96	204	300	396	504	600	702	792	900	1 002	1 098
WM5	235	0	108	210	306	408	504	612	708	804	912	1 008	1 110
BachLehman	260	0	104	200	306	404	502	604	698	808	902	1 004	1 104
WM4	268	0	91	196	298	395	498	595	698	793	893	1 000	1 097
Lehman94	283	0	94	202	298	399	500	596	700	796	900	1 000	1 097
Sparschu	293	0	105	204	301	404	498	605	702	804	904	1 000	1 105
Lindley	308	0	106	202	304	401	501	604	700	806	902	1 003	1 103
LindleyBis	362	0	97	201	297	400	499	598	701	796	901	997	1 099

Fig. 8.36. Values of MSS for different tunings

See the algorithm in Section 3.3 for computing the MSS of any other tuning.

References

1. Agon, C., Amiot, E., Andreatta, M., *Tiling the line with polynomials*, Proceedings ICMC 2005.
2. Agon, C., Amiot, E., Andreatta, M., Ghisi, D., Mandereau, J., *Z-relation and Homometry in Musical Distributions*, JMM 5 **2**, 2011.
3. Amiot, E., *Why Rhythmic Canons Are Interesting*, in: E. Lluis-Puebla, G. Mazzola and T. Noll (eds.), *Perspectives of Mathematical and Computer-Aided Music Theory, EpOs*, 190-209, Universität Osnabrück, 2004.
4. Amiot, E., *Autosimilar Melodies*, Journal of Mathematics and Music, July, 2 **3**, 2008, pp. 157-180.
5. Amiot, E., *Pour en finir avec* le désir, Revue d'Analyse Musicale XXII, 1991, pp. 87-92.
6. Amiot, E., *Rhythmic canons and Galois theory*, Grazer Math. Ber., 347, 2005, pp. 1-25.
7. Amiot, E., *À propos des canons rythmiques*, Gazette des Mathématiciens, SMF Ed., 106, 2005, pp. 43-67.
8. Amiot, E., *New perspectives on rhythmic canons and the spectral conjecture*, in Special Issue "Tiling Problems in Music," Journal of Mathematics and Music, July, 3 **2**, 2009.
9. Amiot, E., *Can a scale have 14 generators?*, Proceedings of MCM, London, Springer 2015, pp. 349-360.
10. Amiot, E., *David Lewin and Maximally Even Sets*, JMM, 1 **3**, 2007, pp. 157-172.
11. Amiot, E., *Structures, Algorithms, and Algebraic Tools for Rhythmic Canons*, Perspectives of New Music 49 (**2**), 2011, pp. 93-143.
12. Amiot, E. *Viewing Diverse Musical Features in Fourier Space: A Survey*, presented at ICMM Puerto Vallarta 2014, to appear.
13. Amiot, E., Sethares, W., *An Algebra for Periodic Rhythms and Scales*, JMM 5 **3**, 2011.
14. Amiot, E.: Sommes nulles de racines de l'unité, in: Bulletin de l'Union des Professeurs de Spéciales 230, 2010, pp. 30-34.
15. Amiot, E., *The Torii of phases*, Proceedings of MCM, Montreal, Springer 2013.
16. Amiot, E., *Discrete Fourier Transform and Bach's Good Temperament*, Music Theory Online (**2**), 2009.
17. Andreatta, M., *On group-theoretical methods applied to music: some compositional and implementational aspects*, in: E. Lluis-Puebla, G. Mazzola, T. Noll (eds.), *Perspectives of Mathematical and Computer-Aided Music Theory, EpOs*, 122-162, Universität Osnabrück, 2004.
18. Andreatta, M., Agon, C., (guest eds), Special Issue "Tiling Problems in Music," Journal of Mathematics and Music, July, 3 **2**, 2009.

© Springer International Publishing Switzerland 2016
E. Amiot, *Music Through Fourier Space*, Computational Music Science,
DOI 10.1007/978-3-319-45581-5

19. Andreatta, M., *De la conjecture de Minkowski aux canons rythmiques mosaïques*, L'Ouvert, n° 114, p. 51-61, March 2007.

20. Babinet, J. *Babinet's principle* is explained online at http://en.wikipedia.org/wiki/Babinet's_principle.

21. Ballinger, B., Benbernou, F., Gomez, F., O'Rourke, J., Toussaint, G., *The Continuous Hexachordal Theorem*, MCM Conference, New Haven, 2009, pp. 11-21.

22. Bartlette, C.A., *A Study of Harmonic Distance and Its Role in Musical Performance*, PhD diss., Eastman School of Music, 2007.

23. Beauguitte, P., *Transformée de Fourier discrète et structures musicales*, Master's thesis, IRCAM 2011 for ATIAM Master. Available online:
    http://articles.ircam.fr/textes/Beauguitte11a/index.pdf.

24. Cafagna V., Vicinanza D., *Myhill property, CV, well-formedness, winding numbers and all that*, Keynote adress to MaMuX seminar in IRCAM - Paris, May 22, 2004.

25. Callender, C., *Continuous Harmonic Spaces*, Journal of Music Theory Volume 51 **2**, 2007.

26. Callender, C., Quinn, I., Tymoczko, D., *Generalized Voice-Leading Spaces*, Science 320 **5874**, 2008, pp. 346-348.

27. Caure, H., *From covering to tiling modulus p (Modulus p Vuza canons: generalities and resolution of the case $\{0, 1, 2^k\}$ with $p = 2$.)*, JMM, 2016.

28. Carey, N., Clampitt, D., Aspects of Well Formed Scales, *Music Theory Spectrum*, **112**, 1989, pp. 187-206.

29. Chmelnitzki, A., *Some problems related to Singer sets*, available online:
    https://homepages.warwick.ac.uk/~maslan/docs/AtiyahPrizeEssay.pdf

30. Clough, J., Douthett, J., *Maximally even sets*, Journal of Music Theory, **35**, 1991, pp. 93-173.

31. Clough, J., Myerson, G., *Variety and Multiplicity in Diatonic Systems*, Journal of Music Theory, **29:** 1985, pp. 249-70.

32. Clough, J., Myerson, G., *Musical Scales and the Generalized Circle of Fifths*, AMM, **93:9**, 1986, pp. 695-701.

33. Clough, J., Douthett, J., Krantz, R., *Maximally Even Sets: A Discovery in Mathematical Music Theory Is Found to Apply in Physics*, Bridges: Mathematical Connections in Art, Music, and Science, Conference Proceedings, ed. Reza Sarhangi, 2000, pp. 193-200.

34. Cohn, R., *Properties and Generability of Transpositionally Invariant Sets*, Journal of Music Theory, **35:1** 1991, pp. 1-32.

35. Coven, E., Meyerowitz, A. *Tiling the integers with one finite set*, in: *J. Alg.* 212, 1999, pp. 161-174.

36. Davalan, J.P., *Perfect rhythmic tilings*, PNM special issue on rhythmic tilings, 2011.

37. de Bruijn, N.G., *On Number Systems*, Nieuw. Arch. Wisk. 3 **4**, 1956, pp. 15-17.

38. Douthett, J., Krantz, R., *Maximally even sets and configurations: common threads in mathematics, physics, and music*, Journal of Combinatorial Optimization, Springer, 2007.
    Online: http://www.springerlink.com/content/g1228n7t44570442/

39. Fidanza, G., *Canoni ritmici*, tesa di Laurea, U. Pisa, 2008.

40. Jedrzejewski, F., *A simple way to compute Vuza canons*, MaMuX seminar, January 2004, http://www.ircam.fr/equipes/repmus/mamux/.

41. Jedrzejewski, F., Johnson, T., *The Structure of Z-Related Sets*, Proceedings MCM 2013, Montreal, 2013, pp. 128-137.

42. Forte, A., "The Structure of Atonal Music," Yale University Press, 1977 (2nd ed).

43. Fripertinger, H. *Remarks on Rhythmical Canons*, Grazer Math. Ber., 347, 2005, pp. 55-68.

44. Fripertinger, H. *Tiling problems in music theory*, in: E. Lluis-Puebla, G. Mazzola, T. Noll (eds.), *Perspectives of Mathematical and Computer-Aided Music Theory, EpOs*, Universität Osnabrück, 2004, pp. 149-164.

45. Fuglede, H., 1974. *Commuting Self-Adjoint Partial Differential Operators and a Group Theoretic Problem*, J. Func. Anal. 16, pp. 101-121.

46. Gilbert, E., *Polynômes cyclotomiques, canons mosaïques et rythmes k-asymétriques*, mémoire de Master ATIAM, Ircam, May 2007.

47. Hajós, G., *Sur les factorisations des groupes abéliens*, in: *Casopsis Pest. Mat. Fys.* 74, 1954, pp. 157-162.

48. Hall, R., Klinsberg, P., *Asymmetric Rhythms and Tiling Canons*, American Mathematical Monthly, Volume 113, Number 10, December 2006, pp. 887-896.

49. Hanson, H., *Harmonic Materials of Modern Music*. Appleton-Century-Crofts, 1960.

50. Hoffman, J., *On Pitch-Class Set Cartography Relations Between Voice-Leading Spaces and Fourier Spaces*, JMT, 52 **2**, 2008.

51. Johnson, T., *Tiling the Line*, Proceedings of J.I.M., Royan, 2001.

52. Johnson, T., *Permutations of 1234, rhythmic canons, block designs, etc*, Curtat Tunnel et Forde, Lausanne, 2014.

53. Johnson, T., *Tiling in My Music*, Perspectives of New Music 49 **2**, 2011, pp. 10-22.

54. Johnson, T., *Self-Similar Melodies*, Two-Eighteen Press, NY 1996 (2nd ed).

55. Kolountzakis, M. *Translational Tilings of the Integers with Long Periods*, Elec. J. of Combinatorics (10)**1**, R22, 2003.

56. Kolountzakis, M. Matolcsi, M., *Complex Hadamard Matrices And the Spectral Set Conjecture*, Collectanea Mathematica **57**, 2006, pp. 281-291. Draft available online:// http://arxiv.org/abs/math.CA/0411512.

57. Kolountzakis, M. Matolcsi, M., *Algorithms for translational tiling*, in Special Issue "Tiling Problems in Music," Journal of Mathematics and Music, July, 3 **2**, 2009.

58. Krumhansl, C., Kessler, E., *Tracing the Dynamic Changes in Perceived Tonal Organization in a Spatial Representation of Musical Keys*, Psychological Review 89 **4**, 1982, pp. 334-368.

59. Łaba, I., *The spectral set conjecture and multiplicative properties of roots of polynomials*, J. London Math. Soc. 65, 2002, pp. 661-671.

60. Łaba, I., and Konyagin, S., *Spectra of certain types of polynomials and tiling of integers with translates of finite sets*, J. Num. Th. 103 **2**, 2003, pp. 267-280.

61. Lagarias, J., and Wang, Y. *Tiling the line with translates of one tile*, in: *Inv. Math.* 124, 1996, pp. 341-365.

62. Lewin, D., *Intervallic Relations Between Two Collections of Notes*, JMT **3**, 1959.

63. Lewin, D., *Special Cases of the Interval Function Between Pitch-Class Sets X and Y*, JMT, 42 **2** 2001, pp. 1-29

64. Mandereau, J., Ghisi, D., Amiot, E., Andreatta, M., Agon, C., *Discrete Phase Retrieval in Musical Distributions*, JMM, **5**, 2011.

65. Mazzola, G., *The Topos of Music*, Birkhäuser, Basel, 2003.

66. Mazzola, G., *Gruppen und Kategorien in der Musik: Entwurf einer mathematischen Musiktheorie*, Heldermann, Lemgo 1985, pp. 11-12.

67. Milne, A., Bulger, D., Herff, S., Sethares, W., *Perfect Balance: A Novel Principle for the Construction of Musical Scales and Meters*, in Proceedings of MCM 5th international conference, London (Springer) 2015, pp. 97-108.

68. Milne, A., Carlé, M., Sethares, W., Noll, T., Holland, S., *Scratching the Scale Labyrinth*, Proceedings MCM 2011 Paris, Springer, 2001, pp. 180-195.

69. Noll, T., Amiot, E., Andreatta, M., *Fourier Oracles for Computer-Aided Improvisation*, Proceedings of the ICMC: Computer Music Conference. Tulane University, New Orleans, 2006. Available online:
`http://architexte.ircam.fr/textes/Amiot06a/index.pdf`

70. Noll, T., Carle, M., *Fourier scratching: SOUNDING CODE*, Presented at the SuperCollider conference, Berlin 2010. Available online:
`http://supercollider2010.de/images/papers/`
`fourier-scratching.pdf.`

71. Perle, G., *Berg's Master Array of the Interval Cycles*, Musical Quarterly 63 **1**, January 2007, pp. 1-30

72. I. Quinn, *General Equal-Tempered Harmony*, Perspectives of New Music 44 **2** - 45 **1**, 2006-2007.

73. Rahn, J., Amiot, E., eds, Perspectives of New Music, special issue 49 **2** on Tiling Rhythmic Canons, 2011.

74. Rahn, J., *Basic Atonal Theory*, New York, Longman, 1980.

75. Rosenblatt, J., Seymour, P.D., *The Structure of Homometric Sets*, SIAM. J. on Algebraic and Discrete Methods Volume 3, Issue 3, 1982.

76. Rosenblatt, J., *Phase Retrieval*, Communications in Mathematical Physics 95, 317-343, 1984.

77. Sands, A.D., *The Factorization of Abelian Groups*, Quart. J. Math. Oxford, **10**, 1962.

78. Schramm, W., *The Fourier transform of functions of the greatest common divisor*, Electronic Journal of Combinatorial Number Theory A50 8 **1**, 2008. Available online:
`http://www.emis.de/journals/INTEGERS/papers/i50/i50.pdf`

79. Sethares, W., "Tuning, Timbre, Spectrum, Scale," Springer, 2013.

80. Singer, J., *A theorem in finite projective geometry and some applications to number theory*, Trans. Amer. Math. Soc., 43, 1938, pp. 377-385.

81. Szabó, S., *A type of factorization of finite abelian groups*, Discrete Math. 54, 1985, pp. 121-124.

82. Tao, T., *Fuglede's conjecture is false in 5 and higher dimensions*, Mathematical Research Letters 11 **2**, July 2003. Available online:
`http://arxiv.org/abs/math.CO/0306134.`

83. Tao, T., *An uncertainty principle for cyclic groups of prime order*, Math. Res. Lett., **12**, 2005, pp. 121-127.

84. Taruskin, R., *Catching up with Rimsky-Korsakov*, Music Theory Spectrum 33 **2**, 2011, pp. 169-85.

85. Terhardt, E., *Pitch, consonance, and harmony*, Journal of the Acoustical Society of America 55 **5**, 1974, pp. 1061-1069.

86. Tijdeman, R., *Decomposition of the Integers as a direct sum of two subsets*, in: Séminaire de théorie des nombres de Paris, 3rd ed., Cambridge University Press, 1995, pp. 261-276.

87. Toussaint, G., *The Geometry of Musical Rhythm*, Chapman and Hall/CRC, January 2013.

88. Tymoczko, D., *Set-class Similarity, Voice Leading, and the Fourier Transform*, Journal of Music Theory, 52 **2** 2008, pp. 251-272.

89. Tymoczko, D., *Three conceptions of musical distance*, Proceedings of MCM, Yale, Springer, 2009, pp. 258-272.

90. Tymoczko, D., *Geometrical Methods in Recent Music Theory*, MTO 16 **1**, 2010. Online:
`http://www.mtosmt.org/issues/mto.10.16.1/mto.10.16.1.`
`tymoczko.html`

91. Tymoczko, D., "A Geometry of Music," Oxford University Press, 2008, pp. 102 and others.

92. Tymoczko, D., *Colloquy: Stravinsky and the Octatonic: Octatonicism Reconsidered Again*, Music Theory Spectrum 25 **1**, 2003, pp. 185-202.
93. Van der Toorn, *Colloquy: Stravinsky and the Octatonic: The Sounds of Stravinsky*, Music Theory Spectrum 25 **1**, 2003, pp. 167-85.
94. Vuza, D.T., *Supplementary Sets and Regular Complementary Unending Canons*, in four parts in *Canons. Persp. of New Music*, 1991-1992: no 29 **2** pp. 22-49; 30 **1**, pp. 184-207; 30 **2**, pp. 102-125; 31 **1**, pp. 270-305.
95. Wild, J., *Tessellating the chromatic*, Perspectives of New Music, 2002.
96. Yust, J., *Schubert's harmonic language and Fourier phase space*, JMT **59**, pp. 121-181 (2015).
97. Yust, J., *Restoring the Structural Status of Keys through DFT Phase Space*, to appear in Proceedings of ICMM, Puerto Vallarta, Springer, 2014.
98. Yust, J., *Applications of DFT to the theory of twentieth-century harmony*, Proceedings of MCM, London, 2015, Springer, pp. 207-218.
99. Yust, J., *Analysis of Twentieth-Century Music Using the Fourier Transform*, Music Theory Society of New York State, Binghamton, April 2015.
100. Yust, J., *Special Collections: Renewing Set Theory*, to appear in JMT, 2016.

Index

© Springer International Publishing Switzerland 2016
E. Amiot, *Music Through Fourier Space*, Computational Music Science,
DOI 10.1007/978-3-319-45581-5

Printed in the United States
By Bookmasters